Columba Sara Evelyn (Ed.)

Nokia 5500 Sport

Columba Sara Evelyn (Ed.)

Nokia 5500 Sport

S60 (software platform), Binary file, Speech synthesis

Fec Publishing

Imprint

All parts of this book are extracted from Wikipedia, the free encyclopedia (www.wikipedia.org).

You can get detailed informations about the authors of this collection of articles at the end of this book. The editors (Ed.) of this book are no authors. They have not modified or extended the original texts.

Pictures published in this book can be under different licences than the GNU Free Documentation License. You can get detailed informations about the authors and licences of pictures at the end of this book.

The content of this book was generated collaboratively by volunteers. Please be advised that nothing found here has necessarily been reviewed by people with the expertise required to provide you with complete, accurate or reliable information. Some information in this book maybe misleading or wrong. The Publisher does not guarantee the validity of the information found here. If you need specific advice (f.e. in fields of medical, legal, financial, or risk management questions) please contact a professional who is licensed or knowledgeable in that area.

Cover image: www.ingimage.com
Concerning the licence of the cover image please contact ingimage.

Publisher:
Fec Publishing is a trademark of
International Book Market Service Ltd., 17 Rue Meldrum, Beau Bassin, 1713-01 Mauritius
Email: info@bookmarketservice.com
Website: www.bookmarketservice.com

Published in 2012

Printed in: U.S.A., U.K., Germany. This book was not produced in Mauritius.

ISBN: 978-620-0-44034-1

Contents

Articles

Nokia_5500_Sport 1
S60_(software_platform) 5
Binary_file 13
Speech_synthesis 15
Motion_detector 26
Office_suite 28
List_of_Nokia_products 30
Smartphone 58
Symbian 83

References

Article Sources and Contributors 102
Image Sources, Licenses and Contributors 104

Nokia_5500_Sport

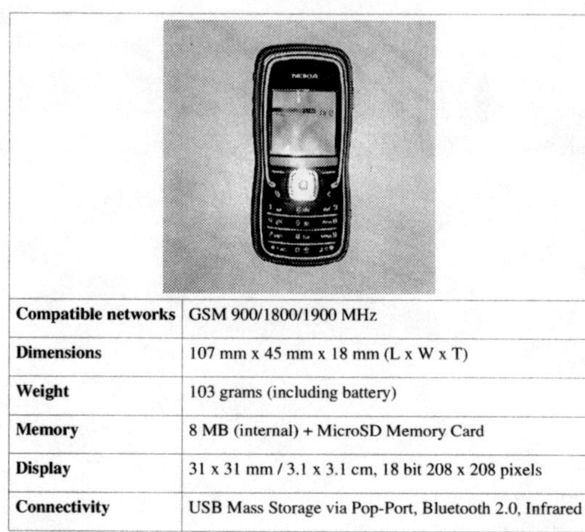

Compatible networks	GSM 900/1800/1900 MHz
Dimensions	107 mm x 45 mm x 18 mm (L x W x T)
Weight	103 grams (including battery)
Memory	8 MB (internal) + MicroSD Memory Card
Display	31 x 31 mm / 3.1 x 3.1 cm, 18 bit 208 x 208 pixels
Connectivity	USB Mass Storage via Pop-Port, Bluetooth 2.0, Infrared

Nokia 5500 Sport is one of the third edition, Nokia S60 platform (S60v3) mobile phones with version 9.1 of the Symbian operating system. It is *not* binary compatible with software compiled for earlier versions of the Symbian operating system. This was the first Nokia handset ever to feature text to speech and motion sensor features.

In addition to Contacts, Messaging, Calendar, and Gallery applications, the Nokia 5500 also includes a view-only Office suite and PDF Viewer.

Differentiating features include:

- Built-in 3D accelerometer: allows phone to act as pedometer/step counter.
- Swap key for quick one-key switching between phone, music, and sports modes.
- Stainless steel body: Built to resist knocks, dust and water splashes.
- Text to speech: Software to read aloud SMS and exercise data.
- Sports tracking: plans, records, and schedules workout sessions.
- Music: Unique tap commands for playing MP3 files.
- Integrated 2-megapixel camera with 4x digital zoom.

Nokia 5500 Sport Music / XpressMusic

Nokia 5500 Sport Music or **XpressMusic** is the music variant model of the Nokia 5500 Sport Model in different color scheme. Its sales package includes added accessories, such as a higher capacity memory card, bike mount and shoulder bag. Apart from that its identical hardware wise to the Nokia 5500 Sport Edition.

Specifications sheet

Feature	Specification
Form factor	Bar / monoblock
Weight	103 grams (including battery)
Dimensions	107 mm x 45 mm x 18 mm (L x W x T)
Volume	77 cc
Talk time	4 hours
Standby time	270 hours (Up to 10 days)
Battery	Li-Ion BL-5B 860 mAh
Operating System	Symbian OS (9.1) + Series 60 v3
GSM frequencies	GSM 900/1800/1900 MHz
GPRS	Yes, GPRS multislot class 10, up to 62.4 kbit/s
EDGE (EGPRS)	Yes, EDGE multislot class 10, up to 236.8 kbit/s
3G	No
UMTS/WCDMA	No
WLAN/WiFi	No
Screen Size	31 x 31 mm / 3.1 x 3.1 cm
Screen Resolution	208 x 208 pixels
Flash Light Torch	Yes
Camera	2.0 megapixel, 4x digital zoom (1600 x 1200, 1152 x 864, 640 x 480 pixels)
Video recording	Yes, QCIF (176 x 144 pixels), sub-QCIF (128 x 96 pixels)
Voice recording	Yes
Text to Speech	Yes, SMS and exercise data read out loud
Multimedia Messaging	Yes
Video calls	No
Push to talk	Yes (Push to Talk over Cellular - PoC)
Java support	Yes, MIDP 2.0
Built-in memory	8 MB
Memory card slot	Yes, microSD (Expandable up to 2 GB[1])
Hot Swappable Memory card slot	No
Bluetooth	Yes, Bluetooth 2.0 + EDR (Enhanced Data Rate)
Infrared	Yes
USB Mass Storage class support	Yes, via Pop-Port
Data cable support	Yes
Browser	WAP 2.0 XHTML / HTML. Comes standard with a full Web browser [2]
Email	Yes
Music player	Yes, stereo, MP3, AAC, AAC, eAAC and WMA playback support
Radio	Yes, FM Stereo with Visual Radio client
Video Player	Yes

Stereo Speakers	Yes
Ringtones	Yes, Monophonic, Polyphonic, True Tones and MP3
Vibrate	Yes
Handsfree Speaker	Yes
Offline/Flight mode	Yes
Released	Q3 2006
Synchronization	PC: Nokia PC Suite, Over the Air: SyncML

Known issues

Some users have complained on the Nokia Support Discussions Board that the Nokia 5500 keypad gets unglued only after a few weeks of light usage, however later release models appear to have had this problem fixed.[3] [4]

See also

- List of Nokia products
- Nokia Series 60
- Smartphone
- Symbian

References

[1] Nokia Accessories Page for 2 GB MicroSD Card (http://europe.nokia.com/accessorieslink?s=MU-37)
[2] http://www.nokia.com/browser
[3] Nokia Support Discussions - Keypad Peeling Discussion 1 (http://discussions.europe.nokia.com/discussions/board/message?board.id=smartphones&message.id=22330)
[4] Nokia Support Discussions - Keypad Peeling Discussion 2 (http://discussions.europe.nokia.com/discussions/board/message?board.id=smartphones&message.id=15812&page=6)

External links

- Nokia 5500 Sport Official Product page (http://europe.nokia.com/5500)
- Nokia 5500 Sport - Device Details (http://www.forum.nokia.com/devices/5500)
- Nokia 5500 Sport - Resource Information (http://www.forum.nokia.com/info/sw.nokia.com/id/ed555638-8299-4f9a-8b5e-a7973912308e/5500.html)
- Nokia 5500 Sport - Phone support (http://europe.nokia.com/A4164325)
- Nokia 5500 Sport XpressMusic - Promotional Campaign (http://www.nokia.com/NOKIA_COM_1/Press/pr_5500music/n5500music.html)

Reviews, photos and videos

- Nokia 5500 Sport - International review roundup by global review aggregator alaTEST (http://alatest.com/Smartphones/Nokia+5500+Sport/pro-reviews/)
- Nokia 5500 Sport XpressMusic - Promotional HighRes Images (http://www.nokia.com/NOKIA_COM_1/Press/pr_5500music/images/product_images_set.zip)
- Nokia 5500 Sport - Review by Mobile-Review (http://www.mobile-review.com/review/nokia-5500-en.shtml)
- Nokia 5500 Sport - Review by GSM Arena (http://www.gsmarena.com/nokia_5500_sport-review-112.php), Video (http://www.gsmarena.com/nokia_5500_sport-video-1557.php)
- Nokia 5500 Sport - Review by All About Symbian (http://www.allaboutsymbian.com/reviews/item/Nokia_5500_Sport_Review1.php), Videocast (http://www.allaboutsymbian.com/media/item/Videocast_5Review_of_the_Nokia_5500.php)
- Nokia 5500 Sport - Review by MobileBurn.com (http://www.mobileburn.com/review.jsp?Id=2876&Page=1#p1s1)
- Nokia 5500 Sport - Review by iMobile.com.au (http://www.imobile.com.au/phonereviews/default.asp?ID=reviewsdec0611)
- Nokia 5500 Sport - CNET Reviews: CNET Australia (http://www.cnet.com.au/mobilephones/phones/0,239025953,339272618,00.htm), CNET U.K. (http://reviews.cnet.co.uk/mobiles/0,39030106,49284774,00.htm)
- Nokia 5500 Sport - Review by Pocket Gamer (http://www.pocketgamer.co.uk/r/Mobile/Nokia+5500+Sport/handset_review.asp?c=1988)
- Nokia 5500 Sport - Review by TrustedReviews (http://www.trustedreviews.com/mobile-phones/review/2006/11/12/Nokia-5500-Sport/p1)
- Nokia 5500 Sport - Review by Tech2.com India (http://www.tech2.com/india/reviews/smart-mobile-phones/nokia-5500-sport/3824/0)
- Nokia 5500 Sport - Durability Test by MForum.ru (Videos) (http://www.mforum.ru/phones/tests/037381.htm)
- Nokia 5500 Sport - Comments on Nokia 5500 (Turkish) (http://www.bedavacep.info/03/07/cep-telefonlari/inceleme-nokia-5500/)

S60_(software_platform)

The **S60** Platform (formerly **Series 60 User Interface**) is a software platform for mobile phones that runs on Symbian OS. It was created by Nokia, who made the platform open source and contributed it to the Symbian Foundation. S60 has been used by mobile device manufacturers including Lenovo, LG Electronics, Panasonic, Samsung,[1] , Sendo,[2] and Siemens mobile. Sony co-created the software with Nokia. Symbian (all Symbian products) is the most popular smartphone OS on the market by 37.6% of the sector's total sales, with 111.6m handsets sold in year 2010.[3]

Screenshot of a modern Nokia S60 user interface.

In addition to the manufacturers the community includes:

- Software integration companies such as Sasken, Elektrobit, Teleca, Digia, Mobica, Atelier.tm
- Semiconductor companies Texas Instruments, STMicroelectronics, Broadcom, Sony, Freescale Semiconductor, Samsung Electronics
- Operators such as Vodafone and Orange who develop and provide S60-based mobile applications and services
- Software developers and independent software vendors (ISVs).

S60 consists of a suite of libraries and standard applications, such as telephony, personal information manager (PIM) tools, and Helix-based multimedia players. It is intended to power fully featured modern phones with large colour screens, which are commonly known as smartphones.

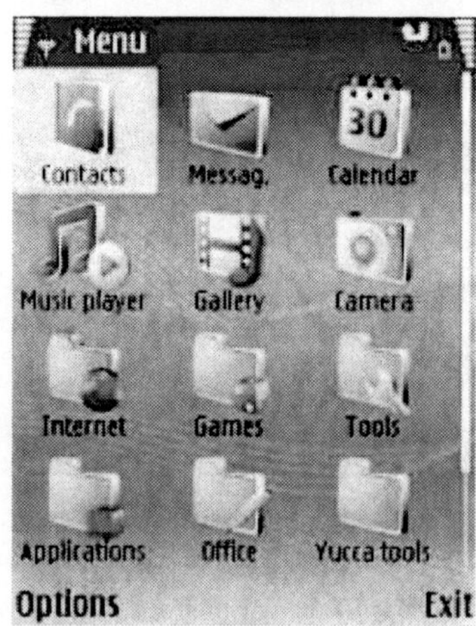

Screenshot of a old Nokia S60 user interface.

The S60 software is a multivendor standard for smartphones that supports application development in Java MIDP, C++, Python[4] and Adobe Flash. Originally, the most distinguishing feature of S60 phones was that they allowed users to install new applications after purchase. Unlike a standard desktop platform, however, the built-in apps are rarely upgraded by the vendor beyond bug fixes. New features are only added to phones while they are being developed rather than after public release. Certain buttons are standardized, such as a menu key, a four way joystick or d-pad, left and right soft keys and a clear key.

S60 editions

There have been four major releases of S60: "Series 60" (2001), "Series 60 Second Edition" (2003), "S60 3rd Edition" (2005) and "S60 5th Edition" (2008).

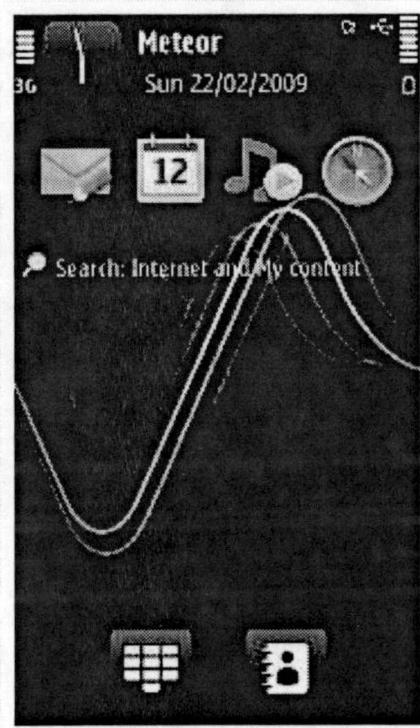

S60 5th edition idle screen. Bottom left "button" brings up a virtual number pad, to compensate for removal of actual numerical keys.

S60 1st Edition

In Series 60 1st Edition, the devices' display resolution was fixed to 176×208.

S60 2nd Edition

Since 2nd Edition Feature Pack 3, Series 60 supports multiple resolutions, i.e. Basic (176×208), and Double (352×416). Nokia N90 was the first S60 device to support a higher resolution (352×416). Some devices, however, have non-standard resolutions, like the Siemens SX1, with 176×220. Nokia 5500 Sport has a 208×208 screen resolution, and the Nokia E90 with its wide 800×352 inner display.

S60 3rd Edition

(S60v3) uses a hardened version of Symbian OS (v9.1), which has mandatory code signing. In S60v3, a user may install only programs that have a certificate from a registered developer, unless the user disables that feature or modify the phone's firmware through third-party hacks that circumvent the mandatory signing restrictions. This makes software written for S60 1st Edition or 2nd Edition not binary-compatible with S60v3.

In 2006, a "Designed for S60 Devices" logo program for developers was launched. The logotype can be used with conforming programs (Symbian or Java)

S60 5th Edition

In October 2008, S60 5th Edition was launched. (Nokia decided to move from 3rd Edition directly to 5th Edition "as a polite gesture to Asian customers",[5] because the number four means bad luck in some Asian cultures). S60 5th Edition runs on Symbian OS version 9.4.[6] The major feature of 5th Edition is support for high-resolution 640×360 touchscreens; before 5th Edition, all S60 devices had a button-based user interface. S60 5th Edition also integrates standard C/C++ APIs and includes Adobe Flash Lite 3.0 with S60-specific ActionScript extensions that give Flash Lite developers access to phone features like contacts, text messaging, sensors and device location information (GPS).

The S60 5th Edition is the last edition of S60. Its assets along with Symbian OS, UIQ and MOAP(S) have been used as a base for Symbian, an open source operating system being developed by the Symbian Foundation. The first edition of Symbian, Symbian^1, uses S60 5th Edition on top of Symbian OS 9.4 as its base.

S60 versions and supported devices

Many devices are capable of running the S60 software platform with the Symbian OS. Devices ranging from the early Nokia 7650 running S60 v0.9 on Symbian OS v6.1,[7] [8] to the latest Samsung i8910 Omnia HD running S60 v5.0 on Symbian OS v9.4.[9] In Symbian^3 the version of the revised platform is v5.2.

The table lists devices carrying each version of S60 as well as the Symbian OS version it is based on.

S60 edition **S60 1st EditionS60 1st Edition,**
Feature Pack 1S60 2nd EditionS60 2nd Edition,
Feature Pack 1S60 2nd Edition,
Feature Pack 2S60 2nd Edition,
Feature Pack 3S60 3rd EditionS60 3rd Edition,
Feature Pack 1S60 3rd Edition,
Feature Pack 2S60 5th Edition
*(Corresponds to Symbian^1)***Symbian^2***"Symbian^3*
(or PR1.0,1.1,1,2,Symbian Anna)"Symbian^3
(or Symbian Belle) S60
version number 0.9 1.2 2.0 2.1 2.6 2.8 3.0 3.1 3.2 5.0 5.2 5.3 [10] Symbian OS
version number 6.1 6.1 7.0s 7.0s 8.0a 8.1a 9.1 9.2 9.3 9.4 9.5 101 [11] Devices[7] [8] [9]

- Nokia 7650
- Nokia 3600
- Nokia 3620
- Nokia 3650
- Nokia 3660
- Nokia N-Gage
- Nokia N-Gage QD
- Sendo X
- Sendo X2 (*canceled*)
- Siemens SX1
- Samsung SGH-D700[12]
- Samsung SGH-D710[12]

- Nokia 6600
- Panasonic X700
- Panasonic X800
- Samsung SGH-D720
- Samsung SGH-D728
- Samsung SGH-D730
- Samsung SGH-Z600

- Nokia 3230
- Nokia 6260
- Nokia 6620
- Nokia 6670
- Nokia 7610

- Lenovo P930
- Nokia 6630
- Nokia 6680
- Nokia 6681

- Nokia 6682
- Nokia N70
- Nokia N72
- Nokia N90
- Nokia 3250
- Nokia 5500 Sport
- Nokia E50
- Nokia E60
- Nokia E61
- Nokia E61i
- Nokia E62
- Nokia E65
- Nokia E70
- Nokia N71
- Nokia N73
- Nokia N75
- Nokia N77
- Nokia N80
- Nokia N91
- Nokia N91 8GB
- Nokia N92
- Nokia N93
- Nokia N93i
- Samsung SGH-i570
- LG KS10
- LG KT610
- LG KT615
- Nokia 5700 XpressMusic
- Nokia 6110 Navigator
- Nokia 6120 Classic
- Nokia 6121 Classic
- Nokia 6124 classic
- Nokia 6290
- Nokia E51
- Nokia E63
- Nokia E66
- Nokia E71
- Nokia E71x
- Nokia E90 Communicator
- Nokia N76
- Nokia N81
- Nokia N81 8GB
- Nokia N82
- Nokia N95
- Nokia N95 8GB
- Samsung SGH-G810
- Samsung SGH-i400

- Samsung SGH-i408
- Samsung SGH-i450
- Samsung SGH-i458
- Samsung SGH-i520
- Samsung SGH-i550
- Samsung SGH-i550w
- Samsung SGH-i560
- Samsung SGH-i568

- Nokia 5320 XpressMusic
- Nokia 5630 XpressMusic
- Nokia 5730 XpressMusic
- Nokia 6210 Navigator
- Nokia 6220 Classic
- Nokia 6650 fold
- Nokia 6710 Navigator
- Nokia 6720 Classic
- Nokia 6730 Classic
- Nokia 6760 Slide
- Nokia 6790 Surge
- Nokia C5-00
- Nokia E5-00
- Nokia E52
- Nokia E55
- Nokia E72
- Nokia E73 Mode
- Nokia E75
- Nokia N78
- Nokia N79
- Nokia N85
- Nokia N86 8MP
- Nokia N96
- Nokia X5
- Samsung GT-i8510 (INNOV8)
- Samsung GT-I7110
- Samsung SGH-L870

- Nokia 5228
- Nokia 5230
- Nokia 5233
- Nokia 5235
- Nokia 5250
- Nokia 5530 XpressMusic
- Nokia 5800 XpressMusic
- Nokia 5800 Navigation Edition
- Nokia C5-03
- Nokia C6-00
- Nokia N97
- Nokia N97 mini

- Nokia X6
- Samsung i8910 Omnia HD[9]
- Sony Ericsson Satio
- Sony Ericsson Vivaz
- Sony Ericsson Vivaz Pro
- DoCoMo F-07B
- DoCoMo F-08B
- DoCoMo F-06B
- DoCoMo SH-07B
- Nokia X7
- Nokia E6
- Nokia 500
- Nokia 702T
- Nokia T7
- Nokia C6-01
- Nokia C7-00
- Nokia E7-00
- Nokia N8
- Nokia 603
- Nokia 700
- Nokia 701

Symbian is now progressing through a period of organisational change to metamorph into an open source software platform project. As an OS, Symbian OS originally provided no user interface (UI), the visual layer that runs atop an operating system. This was implemented separately. Examples of Symbian UIs are MOAP; Series 60; Series 80; Series 90 and UIQ. This separation of UI from underlying OS has created both flexibility and some confusion in the market place. The Nokia purchase of Symbian was brokered with the involvement of the other UI developers and all major user interface layers have been (or have been pledged to be) donated to the open source foundation who will independently own the Symbian operating system. The new Symbian Foundation has announced its intent to unify different Symbian UIs into a single UI based on the S60 platform. (Announcements made in March 2009 indicated this would be the S60 5th edition with feature pack 1).

Symbian Anna

On April 12 2011, Nokia announced Symbian Anna as a software update to the Symbian^3 release. Three new devices (500, X7 and E6) were announced which will have Symbian Anna pre-installed. Symbian Anna will be available as a Software Update for Symbian^3 based devices as well. Most Significant updates that come with "Anna" are

- Portrait QWERTY with split-view data entry
- New Icon Set
- New internet browser with an improved user interface, search-integrated address field, faster navigation and page loading.
- Updated Ovi Maps (search public transport, download full country maps via WLAN or Nokia Ovi Suite, check-in to Facebook, Twitter and Foursquare).
- Java Runtime 2.2, Qt Mobility 1.1 and Qt4.7.

Symbian Belle

On August 24 2011, Nokia announced Symbian Belle as a software update to the Symbian Anna release. Three new devices (603, 700 and 701) [Nokia 600 is cancelled and is replaced with Nokia 603] were announced which will have Symbian Belle pre-installed. Symbian Belle will be available as a Software Update for Symbian Anna based devices as well. Most Significant updates that come with "Belle" are

- Free-form, differently-sized, live widgets
- More homescreens
- Improved status bar
- Dropdown menu
- Modernised navigation
- New apps
- Informative lock screen
- NFC devices
- Visual multitasking

Competitions and alternatives: possible ending

In February 2011, Nokia announced a partnership with Microsoft to integrate Windows Phone 7 as their Primary OS, leaving further Symbian development in question. As a part of this plan, Nokia announced on April 29 2011, to transfer symbian activities to Accenture along with 3000 employees. Timelines for transitions are still not clear. For other alternatives to Symbian, see Symbian.

For historical alternative user interfaces for Symbian (UIQ, series 80, series 90), see history of Symbian.

See also

- Symbian OS
- Android (operating system), an open-source mobile platform by Google
- List of Nokia products
- Location-based services (LBS)
- Maemo, Nokia's Debian Linux-based platform
- Microbrowser
- MOAP, another Symbian-based platform
- Mobile Development
- Mobile game
- Series 40, Nokia's non-Symbian-based platform for mass-market devices.
- UIQ, another Symbian-based platform
- Web Browser for S60

References

[1] *Licensees* (http://www.s60.com/life/thisiss60/S60forbusiness/licensees), S60,

[2] Pakalski, Ingo (21 October 2003). "Symbian-Smartphone von Sendo mit Digitalkamera samt Blitz" (http://www.golem.de/0310/28067.html) (in German). Golem.de. . Retrieved 15 January 2011.

[3] *Gartner: Symbian Is Still the Number One Smartphone Platform [Report* (http://mashable.com/2011/02/10/symbian-number-one-gartner-report/)*], Mashable,*

[4] *Python for S60* (http://www.forum.nokia.com/Resources_and_Information/Tools/Runtimes/Python_for_S60/), Nokia,

[5] S60 5th Edition and the Nokia 5800 XpressMusic are here! (http://blogs.nokia.com/s60blogs/index.php/2008/10/02/s60-5th-edition-and-the-nokia-5800-xpressmusic-are-here/#comment-2107), 3rd comment

[6] "is now closed" (http://www.s60.com/pdfs/getPdf.do?handle=dfb1ff95-77de-4f3b-98a4-4407c519828c-1790878959). S60.com. . Retrieved 2011-03-13.

[7] Forum Nokia Device Specifications for S60 models (http://www.forum.nokia.com/devices/matrix_s60_1.html)(operating system information)

[8] Sony Ericsson Satio Press Release (http://www.sonyericsson.com/cws/corporate/press/pressreleases/pressreleasedetails/satiopressreleasefinal-20090528) (information about S60 version)

[9] Samsung OMNIAHD Dazzles at Mobile World Congress with Its HD Brilliance (http://www.samsung.com/uk/news/newsPreviewRead.do?news_seq=12421)

[10] Screenshot with Nokia 600 (http://upwap.ru/1733517)

[11] Z:\resource\versions\platform.txt (http://upwap.ru/1733462)

[12] "Fast moving phone viruses appear" (http://news.bbc.co.uk/2/hi/technology/4134389.stm). BBC. 30 December 2004. . Retrieved 16 January 2011.

Symbian Belle (http://europe.nokia.com/symbian-belle)

Symbian Belle – the facts, the features and the pictures (http://conversations.nokia.com/2011/08/24/symbian-belle-the-facts-the-features-and-the-pictures/)

External links

- Symbian Anna support forum (http://www.classcep.com/symbian-anna-symbian-belle/)
- Series60 Mobile Phone Themes (http://www.mobilethemez.com/)
- Official links
  - S60 Blogs - The official S60 Blogs by Nokia (http://blogs.s60.com)
  - Forum Nokia - S60 developer site (http://www.forum.nokia.com/series60)
  - Forum.Nokia.com - S60 2nd/3rd Edition: Differences in Features v1.5 (http://www.forum.nokia.com/info/sw.nokia.com/id/571b3f5a-a71a-46d7-958e-79c8081b95c7/S60_2nd_3rd_Ed_Differences_in_Features_v1_5_en.pdf.html)
  - Forum Nokia Russia - S60 developer site (http://www.forum.nokia.ru/Tehnologii_Nokia/Device_Platforms/series_60.html)
  - Open Letter from Nokia CEO to Microsoft CEO (http://conversations.nokia.com/2011/02/11/open-letter-from-ceo-stephen-elop-nokia-and-ceo-steve-ballmer-microsoft/)

Binary_file

A **binary file** is a computer file which may contain any type of data, encoded in binary form for computer storage and processing purposes; for example, computer document files containing formatted text. Many binary file formats contain parts that can be interpreted as text; binary files that contain *only* textual data—without, for example, any formatting information—are called plain text files. In many cases, plain text files are considered to be different from binary files because binary files are made up of more than just plain text. When downloading, a completely functional program without any installer is also often called program binary, or binaries (as opposed to the source code).

```
0000000 0000 0001 0001 1010 0010 0001 0004 012
0000010 0000 0016 0000 0028 0000 0010 0000 002
0000020 0000 0001 0004 0000 0000 0000 0000 000
0000030 0000 0000 0000 0010 0000 0000 0000 020
0000040 0004 8384 0084 c7c8 00c8 4748 0048 e8e
0000050 00e9 6a69 0069 a8a9 00a9 2828 0028 fdf
0000060 00fc 1819 0019 9898 0098 d9d8 00d8 585
0000070 0057 7b7a 007a bab9 00b9 3a3c 003c 888
0000080 8888 8888 8888 8888 288e be88 8888 888
0000090 3b83 5788 8888 8888 7667 778e 8828 888
00000a0 d61f 7abd 8818 8888 467c 585f 8814 818
00000b0 8b06 e8f7 88aa 8388 8b3b 88f3 88bd e98
00000c0 8a18 880c e841 c988 b328 6871 688e 958
00000d0 a948 5862 5884 7e81 3788 1ab4 5a84 3ee
00000e0 3d86 dcb8 5cbb 8888 8888 8888 8888 888
00000f0 8888 8888 8888 8888 8888 8888 8888 000
0000100 0000 0000 0000 0000 0000 0000 0000 000
*
0000130 0000 0000 0000 0000 0000 0000 0000
000013e
```

A hex dump of the 318 byte Wikipedia favicon, or W . The first column numerates the line's starting address, while the * indicates repetition.

Structure

Unlike Text files, there is no special character present in the binary mode files to mark End-of-file. The binary mode files keep track of the end-of-file from number of characters present in the dictionary entry of the file. Binary files are usually thought of as being a sequence of bytes, which means the binary digits (bits) are grouped in eights. Binary files typically contain bytes that are intended to be interpreted as something other than text characters. Compiled computer programs are typical examples; indeed, compiled applications (object files) are sometimes referred to, particularly by programmers, as **binaries**. But binary files can also contain images, sounds, compressed versions of other files, etc. — in short, any type of file content whatsoever.

Some binary files contain headers, blocks of metadata used by a computer program to interpret the data in the file. For example, a GIF file can contain multiple images, and headers are used to identify and describe each block of image data. If a binary file does not contain any headers, it may be called a **flat binary file**. But the presence of headers are also common in plain text files, like email and html files.

Manipulation

To send binary files through certain systems (such as e-mail) that do not allow all data values, they are often translated into a plain text representation (using, for example, Base64). Encoding the data has the disadvantage of increasing the file size during the transfer (for example, using Base64 will increase the file's size by approximately 30%), as well as requiring translation back into binary after receipt. The increased size may be countered by lower-level link compression, as the resulting text data will have about as much less entropy as it has increased size, so the actual data transferred in this scenario would likely be very close to the size of the original binary data. See Binary-to-text encoding for more on this subject.

Microsoft Windows and its standard libraries allow the programmer to specify a parameter indicating if a file is expected to be plain text or binary when opening a file; this affects the standard library calls to read and write from the file in that the system converts between the "line break" character (the ASCII linefeed character) and the line break sequence the operating system expects applications to use in files (the linefeed and carriage return characters in sequence). Unix also allows this, but there are no distinctions between text and binary files for the system to make, as Unix only uses a single line feed character to store a line break in files. This reflects the fact that the distinction between the two types of files is to a certain extent arbitrary.

Viewing

A hex editor or viewer may be used to view file data as a sequence of hexadecimal (or decimal, binary or ASCII character) values for corresponding bytes of a binary file.

If a binary file is opened in a text editor, each group of eight bits will typically be translated as a single character, and you will see a (probably unintelligible) display of textual characters. If the file is opened in some other application, that application will have its own use for each byte: maybe the application will treat each byte as a number and output a stream of numbers between 0 and 255 — or maybe interpret the numbers in the bytes as colors and display the corresponding picture. If the file is itself treated as an executable and run, then the operating system will attempt to interpret the file as a series of instructions in its machine language.

Interpretation

Standards are very important to binary files. For example, a binary file interpreted by the ASCII character set will result in text being displayed. A custom application can interpret the file differently, a byte may be a sound, or a pixel, or even an entire word. Binary itself is meaningless, until such time as an executed algorithm defines what should be done with each bit, byte, word or block. Thus, just examining the binary and attempting to match it against known formats can lead to the wrong conclusion as to what it actually represents. This fact can be used in steganography, where an algorithm interprets a binary data file differently to reveal hidden content. Without the algorithm, it is impossible to tell that hidden content exists.

Binary compatibility

Two files that are binary compatible will have the same pattern of zeros and ones in the data portion of the file. The file header, however, may be different.

The term is used most commonly to state that data files produced by one application are exactly the same as data files produced by another application. For example, some software companies produce applications for Windows and the Macintosh that are binary compatible, which means that a file produced in a Windows environment is interchangeable with a file produced on a Macintosh. This avoids many of the conversion problems caused by importing and exporting data.

See also

- Executable
- Disassembler

External links

- Binary Alphabet [1]

References

[1] http://www.tekmom.com/buzzwords/binaryalphabet.html

Speech_synthesis

Speech synthesis is the artificial production of human speech. A computer system used for this purpose is called a **speech synthesizer**, and can be implemented in software or hardware. A **text-to-speech (TTS)** system converts normal language text into speech; other systems render symbolic linguistic representations like phonetic transcriptions into speech.[1]

Stephen Hawking is one of the most famous people using speech synthesis to communicate

Synthesized speech can be created by concatenating pieces of recorded speech that are stored in a database. Systems differ in the size of the stored speech units; a system that stores phones or diphones provides the largest output range, but may lack clarity. For specific usage domains, the storage of entire words or sentences allows for high-quality output. Alternatively, a synthesizer can incorporate a model of the vocal tract and other human voice characteristics to create a completely "synthetic" voice output.[2]

The quality of a speech synthesizer is judged by its similarity to the human voice and by its ability to be understood. An intelligible text-to-speech program allows people with visual impairments or reading disabilities to listen to written works on a home computer. Many computer operating systems have included speech synthesizers since the early 1980s.

Overview of text processing

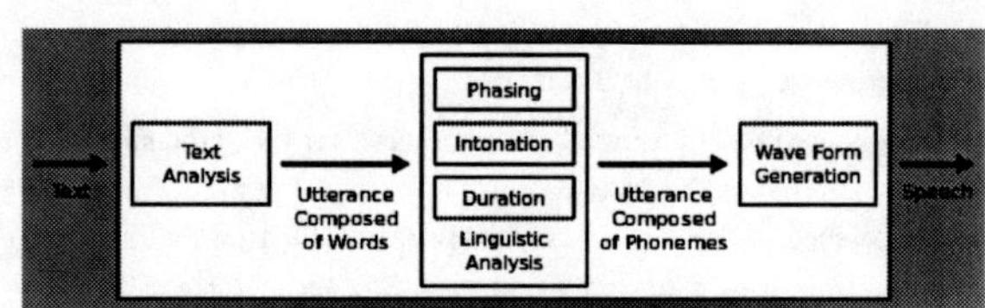

Overview of a typical TTS system

A text-to-speech system (or "engine") is composed of two parts:[3] a front-end and a back-end. The front-end has two major tasks. First, it converts raw text containing symbols like numbers and abbreviations into the equivalent of written-out words. This process is often called *text normalization*, *pre-processing*, or *tokenization*. The front-end then assigns phonetic transcriptions to each word, and divides and marks the text into prosodic units, like phrases, clauses, and sentences. The process of assigning phonetic transcriptions to words is called *text-to-phoneme* or *grapheme-to-phoneme* conversion. Phonetic transcriptions and prosody information together make up the symbolic linguistic representation that is output by the front-end. The back-end—often referred to as the *synthesizer*—then converts the symbolic linguistic representation into sound. In certain systems, this part includes the computation of the *target prosody* (pitch contour, phoneme durations),[4] which is then imposed on the output speech.

History

Long before electronic signal processing was invented, there were those who tried to build machines to create human speech. Some early legends of the existence of "speaking heads" involved Gerbert of Aurillac (d. 1003 AD), Albertus Magnus (1198–1280), and Roger Bacon (1214–1294).

In 1779, the Danish scientist Christian Kratzenstein, working at the Russian Academy of Sciences, built models of the human vocal tract that could produce the five long vowel sounds (in International Phonetic Alphabet notation, they are [aː], [eː], [iː], [oː] and [uː]).[5] This was followed by the bellows-operated "acoustic-mechanical speech machine" by Wolfgang von Kempelen of Vienna, Austria, described in a 1791 paper.[6] This machine added models of the tongue and lips, enabling it to produce consonants as well as vowels. In 1837, Charles Wheatstone produced a "speaking machine" based on von Kempelen's design, and in 1857, M. Faber built the "Euphonia". Wheatstone's design was resurrected in 1923 by Paget.[7]

In the 1930s, Bell Labs developed the VOCODER, a keyboard-operated electronic speech analyzer and synthesizer that was said to be clearly intelligible. Homer Dudley refined this device into the VODER, which he exhibited at the 1939 New York World's Fair.

The Pattern playback was built by Dr. Franklin S. Cooper and his colleagues at Haskins Laboratories in the late 1940s and completed in 1950. There were several different versions of this hardware device but only one currently survives. The machine converts pictures of the acoustic patterns of speech in the form of a spectrogram back into sound. Using this device, Alvin Liberman and colleagues were able to discover acoustic cues for the perception of phonetic segments (consonants and vowels).

Dominant systems in the 1980s and 1990s were the MITalk system, based largely on the work of Dennis Klatt at MIT, and the Bell Labs system;[8] the latter was one of the first multilingual language-independent systems, making extensive use of Natural Language Processing methods.

Early electronic speech synthesizers sounded robotic and were often barely intelligible. The quality of synthesized speech has steadily improved, but output from contemporary speech synthesis systems is still clearly distinguishable from actual human speech.

As the cost-performance ratio causes speech synthesizers to become cheaper and more accessible to the people, more people will benefit from the use of text-to-speech programs.[9]

Electronic devices

The first computer-based speech synthesis systems were created in the late 1950s, and the first complete text-to-speech system was completed in 1968. In 1961, physicist John Larry Kelly, Jr and colleague Louis GerstmanLambert, Bruce (March 21, 1992). "Louis Gerstman, 61, a Specialist In Speech Disorders and Processes" [10]. *New York Times.* used an IBM 704 computer to synthesize speech, an event among the most prominent in the history of Bell Labs. Kelly's voice recorder synthesizer (vocoder) recreated the song "Daisy Bell", with musical accompaniment from Max Mathews. Coincidentally, Arthur C. Clarke was visiting his friend and colleague John Pierce at the Bell Labs Murray Hill facility. Clarke was so impressed by the demonstration that he used it in the climactic scene of his screenplay for his novel *2001: A Space Odyssey*,Arthur C. Clarke Biography [11] at the Wayback Machine (*archived December 11, 1997*) where the HAL 9000 computer sings the same song as it is being put to sleep by astronaut Dave Bowman."Where "HAL" First Spoke (Bell Labs Speech Synthesis website)" [12]. Bell Labs. Retrieved 2010-02-17. Despite the success of purely electronic speech synthesis, research is still being conducted into mechanical speech synthesizers.Anthropomorphic Talking Robot Waseda-Talker Series [13]Handheld electronics featuring speech synthesis began emerging in the 1970s. One of the first was the Telesensory Systems Inc. (TSI) *Speech+* portable calculator for the blind in 1976.TSI Speech+ & other speaking calculators [14] Gevaryahu, Jonathan, "TSI S14001A Speech Synthesizer LSI Integrated Circuit Guide" [15] Other devices were produced primarily for educational purposes, such as Speak & Spell, produced by Texas InstrumentsBreslow, et al.

United States Patent 4326710: "Talking electronic game" [16] April 27, 1982 in 1978. Fidelity released a speaking version of its electronic chess computer in 1979.Voice Chess Challenger [17] The first video game to feature speech synthesis was the 1980 shoot 'em up arcade game, *Stratovox*, from Sun Electronics.Gaming's Most Important Evolutions [18], GamesRadar Another early example was the arcade version of *Bezerk*, released that same year. The first multi-player electronic game using voice synthesis was *Milton* from Milton Bradley Company, which produced the device in 1980.

Synthesizer technologies

The most important qualities of a speech synthesis system are *naturalness* and *intelligibility*. Naturalness describes how closely the output sounds like human speech, while intelligibility is the ease with which the output is understood. The ideal speech synthesizer is both natural and intelligible. Speech synthesis systems usually try to maximize both characteristics.

The two primary technologies for generating synthetic speech waveforms are *concatenative synthesis* and *formant synthesis*. Each technology has strengths and weaknesses, and the intended uses of a synthesis system will typically determine which approach is used.

Concatenative synthesis

Concatenative synthesis is based on the concatenation (or stringing together) of segments of recorded speech. Generally, concatenative synthesis produces the most natural-sounding synthesized speech. However, differences between natural variations in speech and the nature of the automated techniques for segmenting the waveforms sometimes result in audible glitches in the output. There are three main sub-types of concatenative synthesis.

Unit selection synthesis

Unit selection synthesis uses large databases of recorded speech. During database creation, each recorded utterance is segmented into some or all of the following: individual phones, diphones, half-phones, syllables, morphemes, words, phrases, and sentences. Typically, the division into segments is done using a specially modified speech recognizer set to a "forced alignment" mode with some manual correction afterward, using visual representations such as the waveform and spectrogram.[19] An index of the units in the speech database is then created based on the segmentation and acoustic parameters like the fundamental frequency (pitch), duration, position in the syllable, and neighboring phones. At run time, the desired target utterance is created by determining the best chain of candidate units from the database (unit selection). This process is typically achieved using a specially weighted decision tree.

Unit selection provides the greatest naturalness, because it applies only a small amount of digital signal processing (DSP) to the recorded speech. DSP often makes recorded speech sound less natural, although some systems use a small amount of signal processing at the point of concatenation to smooth the waveform. The output from the best unit-selection systems is often indistinguishable from real human voices, especially in contexts for which the TTS system has been tuned. However, maximum naturalness typically require unit-selection speech databases to be very large, in some systems ranging into the gigabytes of recorded data, representing dozens of hours of speech.[20] Also, unit selection algorithms have been known to select segments from a place that results in less than ideal synthesis (e.g. minor words become unclear) even when a better choice exists in the database.[21] Recently, researchers have proposed various automated methods to detect unnatural segments in unit-selection speech synthesis systems.[22]

Diphone synthesis

Diphone synthesis uses a minimal speech database containing all the diphones (sound-to-sound transitions) occurring in a language. The number of diphones depends on the phonotactics of the language: for example, Spanish has about 800 diphones, and German about 2500. In diphone synthesis, only one example of each diphone is contained in the speech database. At runtime, the target prosody of a sentence is superimposed on these minimal units by means of digital signal processing techniques such as linear predictive coding, PSOLA[23] or MBROLA.[24] Diphone synthesis suffers from the sonic glitches of concatenative synthesis and the robotic-sounding nature of formant synthesis, and has few of the advantages of either approach other than small size. As such, its use in commercial applications is declining, although it continues to be used in research because there are a number of freely available software implementations.

Domain-specific synthesis

Domain-specific synthesis concatenates prerecorded words and phrases to create complete utterances. It is used in applications where the variety of texts the system will output is limited to a particular domain, like transit schedule announcements or weather reports.[25] The technology is very simple to implement, and has been in commercial use for a long time, in devices like talking clocks and calculators. The level of naturalness of these systems can be very high because the variety of sentence types is limited, and they closely match the prosody and intonation of the original recordings.

Because these systems are limited by the words and phrases in their databases, they are not general-purpose and can only synthesize the combinations of words and phrases with which they have been preprogrammed. The blending of words within naturally spoken language however can still cause problems unless the many variations are taken into account. For example, in non-rhotic dialects of English the *"r"* in words like *"clear"* /ˈklɪə/ is usually only pronounced when the following word has a vowel as its first letter (e.g. *"clear out"* is realized as /ˌklɪərˈʌʊt/). Likewise in French, many final consonants become no longer silent if followed by a word that begins with a vowel, an effect called liaison. This alternation cannot be reproduced by a simple word-concatenation system, which would require additional complexity to be context-sensitive.

Formant synthesis

Formant synthesis does not use human speech samples at runtime. Instead, the synthesized speech output is created using additive synthesis and an acoustic model (physical modelling synthesis).[26] Parameters such as fundamental frequency, voicing, and noise levels are varied over time to create a waveform of artificial speech. This method is sometimes called *rules-based synthesis*; however, many concatenative systems also have rules-based components. Many systems based on formant synthesis technology generate artificial, robotic-sounding speech that would never be mistaken for human speech. However, maximum naturalness is not always the goal of a speech synthesis system, and formant synthesis systems have advantages over concatenative systems. Formant-synthesized speech can be reliably intelligible, even at very high speeds, avoiding the acoustic glitches that commonly plague concatenative systems. High-speed synthesized speech is used by the visually impaired to quickly navigate computers using a screen reader. Formant synthesizers are usually smaller programs than concatenative systems because they do not have a database of speech samples. They can therefore be used in embedded systems, where memory and microprocessor power are especially limited. Because formant-based systems have complete control of all aspects of the output speech, a wide variety of prosodies and intonations can be output, conveying not just questions and statements, but a variety of emotions and tones of voice.

Examples of non-real-time but highly accurate intonation control in formant synthesis include the work done in the late 1970s for the Texas Instruments toy Speak & Spell, and in the early 1980s Sega arcade machines[27] and in many Atari, Inc. arcade games[28] using the TMS5220 LPC Chips. Creating proper intonation for these projects was painstaking, and the results have yet to be matched by real-time text-to-speech interfaces.[29]

Articulatory synthesis

Articulatory synthesis refers to computational techniques for synthesizing speech based on models of the human vocal tract and the articulation processes occurring there. The first articulatory synthesizer regularly used for laboratory experiments was developed at Haskins Laboratories in the mid-1970s by Philip Rubin, Tom Baer, and Paul Mermelstein. This synthesizer, known as ASY, was based on vocal tract models developed at Bell Laboratories in the 1960s and 1970s by Paul Mermelstein, Cecil Coker, and colleagues.

Until recently, articulatory synthesis models have not been incorporated into commercial speech synthesis systems. A notable exception is the NeXT-based system originally developed and marketed by Trillium Sound Research, a spin-off company of the University of Calgary, where much of the original research was conducted. Following the demise of the various incarnations of NeXT (started by Steve Jobs in the late 1980s and merged with Apple Computer in 1997), the Trillium software was published under the GNU General Public License, with work continuing as gnuspeech. The system, first marketed in 1994, provides full articulatory-based text-to-speech conversion using a waveguide or transmission-line analog of the human oral and nasal tracts controlled by Carré's "distinctive region model".

HMM-based synthesis

HMM-based synthesis is a synthesis method based on hidden Markov models, also called Statistical Parametric Synthesis. In this system, the frequency spectrum (vocal tract), fundamental frequency (vocal source), and duration (prosody) of speech are modeled simultaneously by HMMs. Speech waveforms are generated from HMMs themselves based on the maximum likelihood criterion.[30]

Sinewave synthesis

Sinewave synthesis is a technique for synthesizing speech by replacing the formants (main bands of energy) with pure tone whistles.[31]

Challenges

Text normalization challenges

The process of normalizing text is rarely straightforward. Texts are full of heteronyms, numbers, and abbreviations that all require expansion into a phonetic representation. There are many spellings in English which are pronounced differently based on context. For example, "My latest project is to learn how to better project my voice" contains two pronunciations of "project".

Most text-to-speech (TTS) systems do not generate semantic representations of their input texts, as processes for doing so are not reliable, well understood, or computationally effective. As a result, various heuristic techniques are used to guess the proper way to disambiguate homographs, like examining neighboring words and using statistics about frequency of occurrence.

Recently TTS systems have begun to use HMMs (discussed above) to generate "parts of speech" to aid in disambiguating homographs. This technique is quite successful for many cases such as whether "read" should be pronounced as "red" implying past tense, or as "reed" implying present tense. Typical error rates when using HMMs in this fashion are usually below five percent. These techniques also work well for most European languages, although access to required training corpora is frequently difficult in these languages.

Deciding how to convert numbers is another problem that TTS systems have to address. It is a simple programming challenge to convert a number into words (at least in English), like "1325" becoming "one thousand three hundred twenty-five." However, numbers occur in many different contexts; "1325" may also be read as "one three two five", "thirteen twenty-five" or "thirteen hundred and twenty five". A TTS system can often infer how to expand a number based on surrounding words, numbers, and punctuation, and sometimes the system provides a way to specify the context if it is ambiguous.[32] Roman numerals can also be read differently depending on context. For example "Henry VIII" reads as "Henry the Eighth", while "Chapter VIII" reads as "Chapter Eight".

Similarly, abbreviations can be ambiguous. For example, the abbreviation "in" for "inches" must be differentiated from the word "in", and the address "12 St John St." uses the same abbreviation for both "Saint" and "Street". TTS systems with intelligent front ends can make educated guesses about ambiguous abbreviations, while others provide the same result in all cases, resulting in nonsensical (and sometimes comical) outputs, such as "co-operation" being rendered as "company operation".

Text-to-phoneme challenges

Speech synthesis systems use two basic approaches to determine the pronunciation of a word based on its spelling, a process which is often called text-to-phoneme or grapheme-to-phoneme conversion (phoneme is the term used by linguists to describe distinctive sounds in a language). The simplest approach to text-to-phoneme conversion is the dictionary-based approach, where a large dictionary containing all the words of a language and their correct pronunciations is stored by the program. Determining the correct pronunciation of each word is a matter of looking up each word in the dictionary and replacing the spelling with the pronunciation specified in the dictionary. The other approach is rule-based, in which pronunciation rules are applied to words to determine their pronunciations based on their spellings. This is similar to the "sounding out", or synthetic phonics, approach to learning reading.

Each approach has advantages and drawbacks. The dictionary-based approach is quick and accurate, but completely fails if it is given a word which is not in its dictionary. As dictionary size grows, so too does the memory space requirements of the synthesis system. On the other hand, the rule-based approach works on any input, but the complexity of the rules grows substantially as the system takes into account irregular spellings or pronunciations. (Consider that the word "of" is very common in English, yet is the only word in which the letter "f" is pronounced [v].) As a result, nearly all speech synthesis systems use a combination of these approaches.

Languages with a phonemic orthography have a very regular writing system, and the prediction of the pronunciation of words based on their spellings is quite successful. Speech synthesis systems for such languages often use the rule-based method extensively, resorting to dictionaries only for those few words, like foreign names and borrowings, whose pronunciations are not obvious from their spellings. On the other hand, speech synthesis systems for languages like English, which have extremely irregular spelling systems, are more likely to rely on dictionaries, and to use rule-based methods only for unusual words, or words that aren't in their dictionaries.

Evaluation challenges

The consistent evaluation of speech synthesis systems may be difficult because of a lack of universally agreed objective evaluation criteria. Different organizations often use different speech data. The quality of speech synthesis systems also depends to a large degree on the quality of the production technique (which may involve analogue or digital recording) and on the facilities used to replay the speech. Evaluating speech synthesis systems has therefore often been compromised by differences between production techniques and replay facilities.

Recently, however, some researchers have started to evaluate speech synthesis systems using a common speech dataset.[33]

Prosodics and emotional content

A study in the journal "**Speech Communication**" by Amy Drahota and colleagues at the University of Portsmouth, UK, reported that listeners to voice recordings could determine, at better than chance levels, whether or not the speaker was smiling.[34] [35] [36] It was suggested that identification of the vocal features that signal emotional content may be used to help make synthesized speech sound more natural.

Dedicated hardware

- Votrax
  - SC-01A (analog formant)
  - SC-02 / SSI-263 / "Artic 263"
- General Instrument SP0256-AL2 (CTS256A-AL2)
- Magnevation SpeakJet (www.speechchips.com TTS256)
- Savage Innovations SoundGin
- National Semiconductor DT1050 Digitalker (Mozer)
- Silicon Systems SSI 263 (analog formant)
- Texas Instruments LPC Speech Chips
  - TMS5110A
  - TMS5200
- Oki Semiconductor
  - ML22825 (ADPCM)
  - ML22573 (HQADPCM)
- Toshiba T6721A
- Philips / Signetics
  - Mullard MEA8000
  - PCF8200
- TextSpeak Embedded TTS-EMHD2 Modules - Standalone - World Languages [37]

Computer operating systems or outlets with speech synthesis

Atari

Arguably, the first speech system integrated into an operating system was the 1400XL/1450XL personal computers designed by Atari, Inc. using the Votrax SC01 chip in 1983. The 1400XL/1450XL computers used a Finite State Machine to enable World English Spelling text-to-speech synthesis.[38] Unfortunately, the 1400XL/1450XL personal computers never shipped in quantity.

The Atari ST computers were sold with "stspeech.tos" on floppy disk.

Apple

The first speech system integrated into an operating system that shipped in quantity was Apple Computer's MacInTalk in 1984. The software was licensed from 3rd party developers Joseph Katz and Mark Barton (later, SoftVoice, Inc.) and was featured during the 1984 introduction of the Macintosh computer. Since the 1980s Macintosh Computers offered text to speech capabilities through The MacinTalk software. In the early 1990s Apple expanded its capabilities offering system wide text-to-speech support. With the introduction of faster PowerPC-based computers they included higher quality voice sampling. Apple also introduced speech recognition into its systems which provided a fluid command set. More recently, Apple has added sample-based voices. Starting as a curiosity, the speech system of Apple Macintosh has evolved into a fully supported program, PlainTalk, for people with vision problems. VoiceOver was for the first time featured in Mac OS X Tiger (10.4). During 10.4 (Tiger) & first releases of 10.5 (Leopard) there was only one standard voice shipping with Mac OS X. Starting with 10.6 (Snow Leopard), the user can choose out of a wide range list of multiple voices. VoiceOver voices feature the taking of realistic-sounding breaths between sentences, as well as improved clarity at high read rates over PlainTalk. Mac OS X also includes say, a command-line based application that converts text to audible speech. The AppleScript Standard Additions includes a say verb that allows a script to use any of the installed voices and to control the pitch, speaking rate and modulation of the spoken text.

The Apple iOS operating system used on the iPhone, iPad and iPod Touch uses VoiceOver speech synthesis for accessibility.[39] Some third party applications also provide speech synthesis to facilitate navigating, reading web pages or translating text.

AmigaOS

The second operating system with advanced speech synthesis capabilities was AmigaOS, introduced in 1985. The voice synthesis was licensed by Commodore International from SoftVoice, Inc., who also developed the original MacinTalk text-to-speech system. It featured a complete system of voice emulation, with both male and female voices and "stress" indicator markers, made possible by advanced features of the Amiga hardware audio chipset.[40] It was divided into a narrator device and a translator library. Amiga Speak Handler featured a text-to-speech translator. AmigaOS considered speech synthesis a virtual hardware device, so the user could even redirect console output to it. Some Amiga programs, such as word processors, made extensive use of the speech system.

Microsoft Windows

Modern Windows desktop systems can use SAPI 4 and SAPI 5 components to support speech synthesis and speech recognition. SAPI 4.0 was available as an optional add-on for Windows 95 and Windows 98. Windows 2000 added Narrator, a text–to–speech utility for people who have visual handicaps. Third-party programs such as CoolSpeech, Textaloud and Ultra Hal can perform various text-to-speech tasks such as reading text aloud from a specified website, email account, text document, the Windows clipboard, the user's keyboard typing, etc. Not all programs can use speech synthesis directly.[41] Some programs can use plug-ins, extensions or add-ons to read text aloud. Third-party programs are available that can read text from the system clipboard.

Microsoft Speech Server is a server-based package for voice synthesis and recognition. It is designed for network use with web applications and call centers.

Text-to-Speech (TTS) refers to the ability of computers to read text aloud. A **TTS Engine** converts written text to a phonemic representation, then converts the phonemic representation to waveforms that can be output as sound. TTS engines with different languages, dialects and specialized vocabularies are available through third-party publishers.[42]

Android

Version 1.6 of Android added support for speech synthesis (TTS).[43]

Internet

Currently, there are a number of applications, plugins and gadgets that can read messages directly from an e-mail client and web pages from a web browser or Google Toolbar such as Text-to-voice which is an add-on to Firefox. Some specialized software can narrate RSS-feeds. On one hand, online RSS-narrators simplify information delivery by allowing users to listen to their favourite news sources and to convert them to podcasts. On the other hand, on-line RSS-readers are available on almost any PC connected to the Internet. Users can download generated audio files to portable devices, e.g. with a help of podcast receiver, and listen to them while walking, jogging or commuting to work.

A growing field in Internet based TTS is web-based assistive technology, e.g. 'Browsealoud' from a UK company and Readspeaker. It can deliver TTS functionality to anyone (for reasons of accessibility, convenience, entertainment or information) with access to a web browser. The non-profit project Pediaphon was created in 2006 to provide a similar web-based TTS interface to the Wikipedia.[44]

Other work is being done in the context of the W3C through the W3C Audio Incubator Group [45] with the involvement of The BBC and Google Inc.

Others

- Some e-book readers, such as the Amazon Kindle, Samsung E6, PocketBook eReader Pro, enTourage eDGe, and the Bebook Neo.
- Some models of Texas Instruments home computers produced in 1979 and 1981 (Texas Instruments TI-99/4 and TI-99/4A) were capable of text-to-phoneme synthesis or reciting complete words and phrases (text-to-dictionary), using a very popular Speech Synthesizer peripheral. TI used a proprietary codec to embed complete spoken phrases into applications, primarily video games.[46]
- IBM's OS/2 Warp 4 included VoiceType, a precursor to IBM ViaVoice.
- Systems that operate on free and open source software systems including Linux are various, and include open-source programs such as the Festival Speech Synthesis System which uses diphone-based synthesis (and can use a limited number of MBROLA voices), and gnuspeech which uses articulatory synthesis[47] from the Free Software Foundation.
- Companies which developed speech synthesis systems but which are no longer in this business include BeST Speech (bought by L&H), Eloquent Technology (bought by SpeechWorks), Lernout & Hauspie (bought by Nuance), SpeechWorks (bought by Nuance), Rhetorical Systems (bought by Nuance).
- GPS Navigation units produced by Garmin, Magellan, TomTom and others use speech synthesis for automobile navigation.

Speech synthesis markup languages

A number of markup languages have been established for the rendition of text as speech in an XML-compliant format. The most recent is Speech Synthesis Markup Language (SSML), which became a W3C recommendation in 2004. Older speech synthesis markup languages include Java Speech Markup Language (JSML) and SABLE. Although each of these was proposed as a standard, none of them has been widely adopted.

Speech synthesis markup languages are distinguished from dialogue markup languages. VoiceXML, for example, includes tags related to speech recognition, dialogue management and touchtone dialing, in addition to text-to-speech markup.

Applications

Speech synthesis has long been a vital assistive technology tool and its application in this area is significant and widespread. It allows environmental barriers to be removed for people with a wide range of disabilities. The longest application has been in the use of screen readers for people with visual impairment, but text-to-speech systems are now commonly used by people with dyslexia and other reading difficulties as well as by pre-literate children. They are also frequently employed to aid those with severe speech impairment usually through a dedicated voice output communication aid.

Speech synthesis techniques are also used in entertainment productions such as games and animations. In 2007, Animo Limited announced the development of a software application package based on its speech synthesis software FineSpeech, explicitly geared towards customers in the entertainment industries, able to generate narration and lines of dialogue according to user specifications.[48] The application reached maturity in 2008, when NEC Biglobe announced a web service that allows users to create phrases from the voices of Code Geass: Lelouch of the Rebellion R2 characters.[49]

In recent years, Text to Speech for disability and handicapped communication aids have become widely deployed in Mass Transit. Companies like TalkingSigns and TextSpeak Systems have pioneered solutions such as TTS for Digital Signage for the Blind [37], that work via standard speakers and also radio receivers (ex: BART in the SF Bay area)

See also

- Text-to-voice — Mozilla Firefox extension
- Microsoft text-to-speech voices
- Loquendo
- CereProc
- Comparison of speech synthesizers
- Articulatory synthesis
- Chinese speech synthesis
- Natural language processing
- Paperless office
- Comparison of screen readers
- Comparison of file readers
- Sinewave synthesis
- Speech processing
- Silent speech interface
- Vocaloid

References

[1] Allen, Jonathan; Hunnicutt, M. Sharon; Klatt, Dennis (1987). *From Text to Speech: The MITalk system. Cambridge University Press. ISBN 0-521-30641-8.*

[2] Rubin, P.; Baer, T.; Mermelstein, P. (1981). "An articulatory synthesizer for perceptual research". *Journal of the Acoustical Society of America* **70** (2): 321–328. doi:10.1121/1.386780.

[3] van Santen, Jan P. H.; Sproat, Richard W.; Olive, Joseph P.; Hirschberg, Julia (1997). *Progress in Speech Synthesis*. Springer. ISBN 0-387-94701-9.

[4] Van Santen, J. (April 1994). "Assignment of segmental duration in text-to-speech synthesis". *Computer Speech & Language* **8** (2): 95–128. doi:10.1006/csla.1994.1005.

[5] History and Development of Speech Synthesis (http://www.acoustics.hut.fi/publications/files/theses/lemmetty_mst/chap2.html), Helsinki University of Technology, Retrieved on November 4, 2006

[6] *Mechanismus der menschlichen Sprache nebst der Beschreibung seiner sprechenden Maschine* ("Mechanism of the human speech with description of its speaking machine," J. B. Degen, Wien). (German)

[7] Mattingly, Ignatius G. (1974). Sebeok, Thomas A.. ed. "Speech synthesis for phonetic and phonological models" (http://www.haskins.yale.edu/Reprints/HL0173.pdf). *Current Trends in Linguistics* (Mouton, The Hague) **12**: 2451–2487. .

[8] Sproat, Richard W. (1997). *Multilingual Text-to-Speech Synthesis: The Bell Labs Approach*. Springer. ISBN 0792380274.

[9] Kurzweil, Raymond (2005). *The Singularity is Near*. Penguin Books. ISBN 0-14-303788-9.

[10] http://www.nytimes.com/1992/03/21/nyregion/louis-gerstman-61-a-specialist-in-speech-disorders-and-processes.html

[11] http://web.archive.org/19971211154551/http://www.lsi.usp.br/~rbianchi/clarke/ACC.Biography.html

[12] http://www.bell-labs.com/news/1997/march/5/2.html

[13] http://www.takanishi.mech.waseda.ac.jp/top/research/voice/index.htm

[14] http://www.vintagecalculators.com/html/speech_.html

[15] http://si.fileburst.com/s14001a_reveng.pdf

[16] http://patft.uspto.gov/netacgi/nph-Parser?Sect2=PTO1&Sect2=HITOFF&p=1&u=%2Fnetahtml%2Fsearch-bool.html&r=1&f=G&l=50&d=PALL&RefSrch=yes&Query=PN%2F4326710

[17] http://www.ismenio.com/chess_fidelity_vcc.html

[18] http://www.gamesradar.com/f/gamings-most-important-evolutions/a-2010100810233l322035/p-2

[19] Alan W. Black, Perfect synthesis for all of the people all of the time. (http://www.cs.cmu.edu/~awb/papers/IEEE2002/allthetime/allthetime.html) IEEE TTS Workshop 2002.

[20] John Kominek and Alan W. Black. (2003). CMU ARCTIC databases for speech synthesis. CMU-LTI-03-177. Language Technologies Institute, School of Computer Science, Carnegie Mellon University.

[21] Julia Zhang. Language Generation and Speech Synthesis in Dialogues for Language Learning (http://groups.csail.mit.edu/sls/publications/2004/zhang_thesis.pdf), masters thesis, Section 5.6 on page 54.

[22] William Yang Wang and Kallirroi Georgila. (2011). Automatic Detection of Unnatural Word-Level Segments in Unit-Selection Speech Synthesis (http://www.cs.cmu.edu/~yww/papers/asru2011.pdf), IEEE ASRU 2011.

[23] Pitch-Synchronous Overlap and Add (PSOLA} Synthesis (http://web.archive.org/20070222180903/http://www.fon.hum.uva.nl/praat/manual/PSOLA.html) at the Wayback Machine (*archived February 22, 2007*)

[24] T. Dutoit, V. Pagel, N. Pierret, F. Bataille, O. van der Vrecken. The MBROLA Project: Towards a set of high quality speech synthesizers of use for non commercial purposes. *ICSLP Proceedings*, 1996.

[25] L.F. Lamel, J.L. Gauvain, B. Prouts, C. Bouhier, R. Boesch. Generation and Synthesis of Broadcast Messages, *Proceedings ESCA-NATO Workshop and Applications of Speech Technology*, September 1993.

[26] Dartmouth College: *Music and Computers* (http://digitalmusics.dartmouth.edu/~book/MATCpages/chap.4/4.4.formant_synth.html), 1993.

[27] Examples include Astro Blaster, Space Fury, and Star Trek: Strategic Operations Simulator

[28] Examples include Star Wars, Firefox, Return of the Jedi, Road Runner, The Empire Strikes Back, Indiana Jones and the Temple of Doom, 720°, Gauntlet, Gauntlet II, A.P.B., Paperboy, RoadBlasters, Vindicators Part II (http://www.arcade-museum.com/game_detail.php?game_id=10319), Escape from the Planet of the Robot Monsters.

[29] John Holmes and Wendy Holmes (2001). *Speech Synthesis and Recognition* (2nd ed.). CRC. ISBN 0-7484-0856-8.

[30] The HMM-based Speech Synthesis System (http://hts.sp.nitech.ac.jp/)

[31] Remez, R.; Rubin, P.; Pisoni, D.; Carrell, T. (22 May 1981). "Speech perception without traditional speech cues" (http://www.bsos.umd.edu/hesp/mwinn/Remez_et_al_1981.pdf). *Science* **212** (4497): 947–949. doi:10.1126/science.7233191. PMID 7233191. .

[32] "Speech synthesis" (http://www.w3.org/TR/speech-synthesis/#S3.1.8). World Wide Web Organization. .

[33] Blizzard Challenge (http://festvox.org/blizzard)

[34] "Smile -and the world can hear you" (http://web.archive.org/web/20080517102201/http://www.port.ac.uk/aboutus/newsandevents/news/title,74220,en.html). University of Portsmouth. January 9, 2008. Archived from the original (http://www.port.ac.uk/aboutus/newsandevents/news/title,74220,en.html) on 2008-05-17. .

[35] "Smile - And The World Can Hear You, Even If You Hide" (http://www.sciencedaily.com/releases/2008/01/080111224745.htm). *Science Daily*. January 2008. .

[36] Drahota, A. (2008). "The vocal communication of different kinds of smile" (http://peer.ccsd.cnrs.fr/docs/00/49/91/97/PDF/PEER_stage2_10.1016%2Fj.specom.2007.10.001.pdf). *Speech Communication* **50** (4): 278–287. doi:10.1016/j.specom.2007.10.001. .

[37] http://www.textspeak.com/oemtts.htm

[38] 1400XL/1450XL Speech Handler External Reference Specification (http://www.atarimuseum.com/ahs_archives/archives/pdf/computers/8bits/1400xlmodem.pdf)

[39] "iPhone: Configuring accessibility features (Including VoiceOver and Zoom)" (http://support.apple.com/kb/ht3577). Apple. . Retrieved 2011-01-29.

[40] Miner, Jay et al. (1991). *Amiga Hardware Reference Manual* (3rd ed.). Addison-Wesley Publishing Company, Inc.. ISBN 0-201-56776-8.

[41] "Accessibility Tutorials for Windows XP: Using Narrator" (http://www.microsoft.com/enable/training/windowsxp/usingnarrator.aspx). Microsoft. 2011-01-29. . Retrieved 2011-01-29.

[42] "How to configure and use Text-to-Speech in Windows XP and in Windows Vista" (http://support.microsoft.com/kb/306902). Microsoft. 2007-05-07. . Retrieved 2010-02-17.

[43] Jean-Michel Trivi (2009-09-23). "An introduction to Text-To-Speech in Android" (http://android-developers.blogspot.com/2009/09/introduction-to-text-to-speech-in.html). Android-developers.blogspot.com. . Retrieved 2010-02-17.

[44] Andreas Bischoff, The Pediaphon - Speech Interface to the free Wikipedia Encyclopedia for Mobile Phones (http://www.dr-bischoff.de/research/pdf/bischoff_pediaphon_uwsi2007_final.pdf), PDA's and MP3-Players, Proceedings of the 18th International Conference on Database and Expert Systems Applications, Pages: 575-579 ISBN 0-7695-2932-1, 2007

[45] http://www.w3.org/2010/04/audio/audio-incubator-charter.html

[46] "Smithsonian Speech Synthesis History Project (SSSHP) 1986-2002" (http://www.mindspring.com/~ssshp/ssshp_cd/ss_home.htm). Mindspring.com. . Retrieved 2010-02-17.

[47] "gnuspeech" (http://www.gnu.org/software/gnuspeech/). Gnu.org. . Retrieved 2010-02-17.

[48] "Speech Synthesis Software for Anime Announced" (http://www.animenewsnetwork.com/news/2007-05-02/speech-synthesis-software). Anime News Network. 2007-05-02. . Retrieved 2010-02-17.

[49] "Code Geass Speech Synthesizer Service Offered in Japan" (http://www.animenewsnetwork.com/news/2008-09-09/code-geass-voice-synthesis-service-offered-in-japan). Animenewsnetwork.com. 2008-09-09. . Retrieved 2010-02-17.

External links

- Text to Speech Synthesis in the Web Browser with JavaScript (http://vimeo.com/12039415)
- Speech synthesis (http://www.dmoz.org/Computers/Speech_Technology/Speech_Synthesis//) at the Open Directory Project
- Text to Voice or Text to Speech Firefox Addon (https://addons.mozilla.org/en-US/firefox/addon/91405/?src=external-wp)
- Dennis Klatt's History of Speech Synthesis (http://www.cs.indiana.edu/rhythmsp/ASA/Contents.html)
- CereProc Text-To-Speech (http://www.cereproc.com)
- IVONA Text-To-Speech (http://www.ivona.com)
- Verbose Text-To-Speech (http://www.nch.com.au/verbose/index.html).
- Text to Speech for Web site (http://www.portalvoz.com)
- RSS to Speech Application and Windows desktop gadget (based on TTS) (http://www.apexoft.com/rss2speech/)
- RSS to Speech Google Desktop gadget (http://desktop.google.com/plugins/i/rsstospeech.html)
- Simulated singing with the singing robot Pavarobotti (http://www.youtube.com/watch?v=CE6zy8aUwtQ) or a description from the BBC on how the robot synthesized the singing (http://www.youtube.com/watch?v=SNqNM6Ccck8).
- Reviews of Popular Speech Synthesizers (http://www.webbie.org.uk/Veli-Pekka/reviews_of_speech_synths.html) by Veli-Pekka Tätilä
- Free online human voice, speaking an entered text. Round trip translation and speech of text, available in many languages. (http://trans121.com)

Motion_detector

A **motion detector** is a device for motion detection. That is, it is a device that contains a physical mechanism or electronic sensor that quantifies motion that can be either integrated with or connected to other devices that alert the user of the presence of a moving object within the field of view. They form a vital component of comprehensive security systems, for both homes and businesses.

A motion detector attached to a garage

Overview

An electronic **motion detector** contains a motion sensor that transforms the detection of motion into an electric signal. This can be achieved by measuring optical or acoustical changes in the field of view. Most motion detectors can detect up to 15 – 25 meters (50–80ft).

A motion detector may be connected to a burglar alarm that is used to alert the home owner or security service after it detects motion. Such a detector may also trigger a red light camera or outdoor lighting.

An occupancy sensor is a motion detector that is integrated with a timing device. It senses when motion has stopped for a specified time period in order to trigger a light extinguishing signal. These devices prevent illumination of unoccupied spaces like public toilets. They are widely used for security purposes.

Sensors

There are basically four types of sensors used in motion detectors spectrum:

Passive infrared sensors (Passive)

Looks for body heat. No energy is emitted from the sensor.

Ultrasonic (active)

Sends out pulses of ultrasonic waves and measures the reflection off a moving object.

Microwave (active)

Sensor sends out microwave pulses and measures the reflection off a moving object. Similar to a police radar gun.

Tomographic Detector (active)

Senses disturbances to radio waves as they travel through an area surrounded by mesh network nodes.

Dual-technology motion detectors

Many modern motion detectors use a combination of different technologies. These dual-technology detectors benefit with each type of sensor, and false alarms are reduced. Placement of the sensors can be strategically mounted so as to lessen the chance of pets activating alarms.

Often, PIR technology will be paired with another model to maximize accuracy and reduce energy usage. PIR draws less energy than microwave detection, and so many sensors are calibrated so that when the PIR sensor is tripped, it activates a microwave sensor. If the latter also picks up an intruder, then the alarm is sounded. As interior motion detectors do not 'see' through windows or walls, motion-sensitive outdoor lighting is often recommended to enhance comprehensive efforts to protect your property.

False alarms are those usually caused by technical errors such as electrical and mechanical failures. Nuisance alarms are system activations not commonly caused by attackers or intruders but rather from wind blown debris, animals, insects and foliage.

Sequencing alarm systems to trip the alert mechanism only when both alarm sensors have been activated will reduce nuisance alarms, but may also cause the probability of detection to decrease.

See also

- Smoke detector
- Heat detector
- Pickup (music technology)
- Motion detection
- Motion controller for video game consoles

External links

- How to build your own motion detector (http://www.jameco.com/Jameco/PressRoom/proximity.html)
- Xandem: Tomographic motion detector company (http://www.xandem.com/motion-detection)

Office_suite

In computing, an **office suite**, sometimes called an **office software suite** or **productivity suite** is a collection of productivity programs intended to be used by knowledge workers. The components are generally distributed together, have a consistent user interface and usually can interact with each other, sometimes in ways that the operating system would not normally allow.

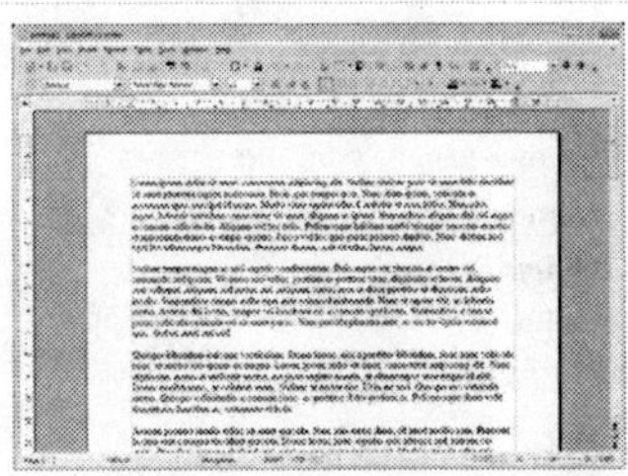

LibreOffice Writer is an example of a word processor.

Typical office suite components

Existing office suites contain wide range of various components. Most typically, the base components include:

- Word processor
- Spreadsheet
- Presentation program

Less common components of office suites include:

- Database
- Graphics suite (raster graphics editor, vector graphics editor, image viewer)
- Desktop publishing software
- Formula editor
- Diagramming software
- Email client
- Communication
- Personal information manager
- Notetaking program
- Groupware
- Project management software
- Web log analysis software

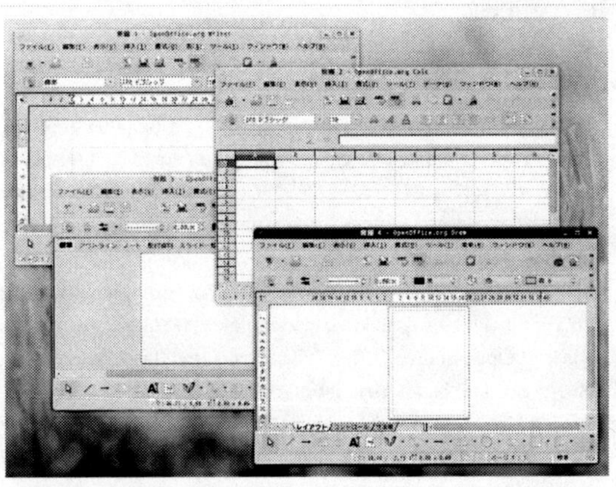

Screen shot for Office Suite OpenOffice.org showing Writer / Calc / Impress / Draw

See also

- List of office suites
- Comparison of office suites
- List of collaborative software
- List of word processors
- Comparison of word processors
- Comparison of spreadsheet software
- Online spreadsheet
- Online office suite
- Office Open XML software
- OpenDocument software
- Web desktop
- Web operating system

References

External links

- DMOZ / Office Suites (http://www.dmoz.org/Computers/Software/Office_Suites/)
- Comparison of online office suite with reviews and cost (http://www.pcmag.com/products/0,,tqs=3CDB422BB1FE1251736923BD8B16980E81F74C38,00.asp?action=defaultadvancedquery&cid=1739&sid=1739&gridtitle=Recent Product Reviews&googlequery=q=&mt823=21132&start=0&sort=articledate&dir=desc&stpdinglp=1)
- Free Web-based Office alternatives (http://www.computerworld.com/action/article.do?command=viewArticleBasic&articleId=9007884)
- Review: Open-Source Office Suites Compared (http://www.informationweek.com/news/software/enterpriseapps/showArticle.jhtml?articleID=212201932&pgno=1&queryText=&isPrev=)
- Review: CNET Office suites (http://download.cnet.com/8301-20_4-10477465-10084490.html)
- Review: PCWorld Office Productivity (http://www.pcworld.com/article/id,143367-page,2-c,downloads/article.html)

List_of_Nokia_products

This list is incomplete.

The following is a **list of products** branded by **Nokia Corporation**. This list concentrates on the modern Nokia products.[1]

Most Nokia phones will show their model number if the user types *#0000# on its keypad in standby mode and unlocked state.

Mobile phones

Phones in **boldface** indicates that the phone is a Symbian, Maemo/MeeGo or Windows Phone-powered smartphone.

S.: status, where

- **P** indicates *under production*,
- **D** is *discontinued*, and
- **U** is *upcoming*.

DCT (Digital Core Technology) **1**, **2**, **3** and **4** are generations of internal Nokia technology.

Classic series – The Mobira series

Phone model	Screen type	Released	S.	Technology	Generation	Form factor
Mobira Senator	Monochrome	1982	D	NMT-450	1G	Boxtype - The first mobile phone
Mobira Talkman 450	Monochrome	1985	D	NMT-450	1G	Boxtype
Mobira Cityman 100	Monochrome	1988	D	ETACS	1G	Brick
Mobira Cityman 150	Monochrome	1989	D	NMT-900	1G	Brick
Mobira Cityman 1320	Monochrome	1987	D	TACS	1G	Brick
Mobira Cityman 900	Monochrome	1987	D	NMT-900	1G	Brick

Original series

Phone model	Screen type	Released	S.	Technology	Generation	Form factor
Nokia 100	Monochrome	1987	D	NMT-900	1G	Candybar
Nokia 101	Monochrome	1992	D	NMT-900	2G	Candybar
Nokia 252	Monochrome	1998	D	AMPS/N-AMPS	Unknown	Candybar
Nokia 282	Monochrome	1998	D	AMPS/N-AMPS	Unknown	Clamshell
Nokia 636	Monochrome	1995	D	AMPS	Unknown	Candybar
Nokia 638	Monochrome	1996	D	AMPS	Unknown	Candybar
Nokia 640	Monochrome	1999	D	NMT-900	Unknown	Candybar
Nokia 650	Monochrome	1999	D	NMT-900	Unknown	Candybar
Nokia 810 (car phone)	Monochrome	2003	D	GSM	Unknown	Candybar
Nokia 918	Monochrome	1996	D	AMPS	Unknown	Candybar
Nokia 1000	Monochrome	1992	D	AMPS	DCT1	Candybar
Nokia 1011	Monochrome	1992	D	GSM	DCT1	Candybar - First GSM phone

Nokia Ringo	Monochrome	1995	D	TACS/ETACS/NMT-900	DCT1	Candybar

1000–9000 series

Nokia 1000 series – Ultrabasic series

The Nokia 1000 series include Nokia's most affordable phones. They are mostly targeted towards developing countries and users who do not require advanced features beyond making calls and SMS text messages, alarm clock, reminders, etc.

Phone model	Screen type	Released	S.	Technology	Generation	Form factor	Ringtone	Images
Nokia 1006	128 × 160 color	2009	P	CDMA		Candybar		
Nokia 1100	96 × 65 Monochrome	2003	D	GSM	DCT4	Candybar	mono	
Nokia 1101	96 × 65 Monochrome	2005	D	GSM	DCT4	Candybar	mono	
Nokia 1108	96 × 65 Monochrome	2005	D	GSM	DCT4+	Candybar	mono	
Nokia 1110	96 × 68 Inverse Monochrome	2005	D	GSM	DCT4	Candybar	MP3	
Nokia 1110i	96 × 68 Monochrome	2006	D	GSM	DCT4+	Candybar	MP3	
Nokia 1112	96 × 68 Monochrome	2006	D	GSM	DCT4+	Candybar	MP3	
Nokia 1200	96 × 68 Monochrome	2007	D	GSM	DCT4	Candybar	MP3	
Nokia 1202	96 × 68 Monochrome	2008	D	GSM	DCT4+	Candybar	MP3	
Nokia 1203	96 × 68 Monochrome	2009	D	GSM	DCT4+	Candybar	MP3	
Nokia 1208	96 × 68 CSTN (color super-twist nematic)	2007	D	GSM	DCT4+	Candybar	MP3	
Nokia 1209	96 × 68 pixels CSTN (color super-twist nematic)	2007	D	GSM	DCT4+	Candybar	MP3	
Nokia 1220	84 × 48 Monochrome	2002	D	TDMA/AMPS	DCT3	Candybar		
Nokia 1221	84 × 48 Monochrome	2002	D	TDMA/AMPS	DCT3	Candybar		
Nokia 1260	84 × 48 Monochrome	2002	D	TDMA/AMPS	DCT3	Candybar		
Nokia 1261	84 × 48 Monochrome	2002	D	TDMA/AMPS	DCT3	Candybar		
Nokia 1280	96 × 68 Monochrome	2010	P	GSM		Candybar		
Nokia 1600	96 × 68 16-bit (65,536) Color	2006	D	GSM	DCT4	Candybar	MP3	
Nokia 1610	Monochrome	1996	D	GSM	DCT1	Candybar		
Nokia 1611	Monochrome	1997	D	GSM	DCT1	Candybar		
Nokia 1616	128 × 160 (65,536) Color	2010	P	GSM		Candybar	MP3	
Nokia 1620	Monochrome	1997	D	GSM	DCT1	Candybar		
Nokia 1630	Monochrome	1998	D	GSM	DCT1	Candybar		
Nokia 1631	Monochrome	1998	D	GSM	DCT1	Candybar		
Nokia 1650	(65,536) Color	2008	D	GSM	DCT4+	Candybar	MP3	

Nokia 1661	128 × 160 (65,536) Color	2008	P	GSM	dct4+	Candybar	MP3	
Nokia 1662	128 × 160 (65,536) Color	2008	P	GSM	dct4+	Candybar	MP3	
Nokia 1680 classic	128 × 160 (65,536) Color	2008	D	GSM	dct4+	Candybar	MP3	
Nokia 1800	128 × 160 (65,536) Color	2010	P	GSM		Candybar	MP3	

Nokia 2000 series – Basic series

Like the 1000 series, the 2000 series are entry-level phones. However, the 2000 series generally contain more advanced features than the 1000 series; many new 2000 series phones feature color screens and some feature cameras, Bluetooth and even A-GPS, GPS such as in the case of the Nokia 2710. The 2000 series slot is between the 1000 and 3000 series phones in terms of features.

Phone model	Screen type	Released	S.	Technology	Generation	Form factor	Camera
Nokia 2010	Monochrome	1994	D	GSM	DCT1	Candybar	None
Nokia 2100	96 × 65 Monochrome	2003	D	GSM	DCT3	Candybar	None
Nokia 2110	Monochrome	1994	D	GSM	DCT1	Candybar	None
Nokia 2110i	Monochrome	1996	D	GSM	DCT1	Candybar	None
Nokia 2112	96 × 65 Monochrome	2004	D	CDMA2000 1x	DCT4	Candybar	None
Nokia 2115	96 × 65 Monochrome	2005	D	CDMA2000 1x	DCT4	Candybar	None
Nokia 2115i	96 × 65 Monochrome	2005	D	CDMA2000 1x	DCT4	Candybar	None
Nokia 2116i	96 × 65 Monochrome	2005	D	CDMA2000 1x	DCT4	Candybar	None
Nokia 2118	96 × 65 Monochrome	2005	D	CDMA2000 1x	DCT4	Candybar	None
Nokia 2120	96 × 65 Monochrome	1994	D	TDMA/AMPS	DCT2	Candybar	None
Nokia 2125i	96 × 65 16-bit (65,536) Color	2005	D	CDMA2000 1x/AMPS	DCT4	Candybar	None
Nokia 2126i	96 × 65 16-bit (65,536) Color	2005	D	CDMA2000 1x	DCT4	Candybar	None
Nokia 2128i	96 × 65 16-bit (65,536) Color	2005	D	CDMA2000 1x	DCT4	Candybar	None
Nokia 2160	Monochrome	1996	D	TDMA/AMPS	DCT1	Candybar	None
Nokia 2160i	Monochrome	1997	D	TDMA/AMPS	DCT1	Candybar	None
Nokia 2168	Monochrome	1997	D	TDMA/AMPS	DCT1	Candybar	None
Nokia 2170	Monochrome	1998	D	CDMA	DCT1	Candybar	None
Nokia 2180	Monochrome	1997	D	CDMA/AMPS	DCT1	Candybar	None
Nokia 2190	Monochrome	1997	D	GSM	DCT1	Candybar	None
Nokia 2220	84 × 48 Monochrome	2003	D	TDMA/AMPS	DCT4	Candybar	None
Nokia 2220 slide	128 × 160 16-bit (65536) color	2010	P		DCT4+	Slide	VGA 0.3 MP
Nokia 2255	128 × 128 (65,536) Color	2005	D	CDMA2000 1x	DCT4	Clamshell	None
Nokia 2260	84 × 48 Monochrome	2003	D	TDMA/AMPS	DCT4	Candybar	None
Nokia 2270	96 × 65 Monochrome	2003	D	CDMA2000 1x	DCT4	Candybar	None
Nokia 2272	96 × 65 Monochrome	2003	D	CDMA2000 1x	DCT4	Candybar	None
Nokia 2280	96 × 65 Monochrome	2003	D	CDMA2000 1x	DCT4	Candybar	None

Nokia 2285	96 × 65 Monochrome	2003	D	CDMA2000 1x	DCT4	Candybar	None
Nokia 2300	96 × 65 Monochrome	2004	D	GSM	DCT4	Candybar	None
Nokia 2310	96 × 68 16-bit (65536) color	2006	D	GSM	DCT4+	Candybar	None
Nokia 2323	96 × 68 16-bit (65536) color	2009	D	GSM	DCT4+	Candybar	None
Nokia 2330	128 × 160 16-bit (65536) color	2009	D	GSM	DCT4+	Candybar	VGA 0.3 MP
Nokia 2366i	128 × 160 16-bit (65536) color	2006	D	CDMA2000 1x	DCT4	Clamshell	None
Nokia 2600	128 × 128 12-bit (4096) Color	2004	D	GSM	DCT4	Candybar	None
Nokia 2600 classic	128 × 160 16-bit (65536) color	2007	D	GSM	DCT4+	Candybar	VGA 0.3 MP
Nokia 2610	128 × 128 16-bit (65536) color	2006	D	GSM	DCT4+	Candybar	None
Nokia 2626	128 × 128 16-bit (65536) color	2007	D	GSM	DCT4+	Candybar	None
Nokia 2630	128 × 160 16-bit (65536) color	2007	D	GSM	DCT4+	Candybar	VGA 0.3 MP
Nokia 2650	128 × 128 12-bit (4096) Color	2004	D	GSM	DCT4	Clamshell	None
Nokia 2651	128 × 128 12-bit (4096) Color	2004	D	GSM	DCT4	Clamshell	None
Nokia 2652	128 × 128 12-bit (4096) Color	2005	D	GSM	DCT4	Clamshell	None
Nokia 2680 slide	128 × 160 16-bit (65536) Color	2008	D	GSM	DCT4+	Slide	VGA 0.3 MP
Nokia 2690	128 × 160 (262144) color	2010	P	GSM	DCT+4	Candybar	VGA 0.3 MP
Nokia 2700 classic	240 × 320 262k Color	2009	P	GSM	BB5.0	Candybar	2.0 MP
Nokia 2710	240 × 320 262k Color	2009	P	GSM	BB5.0	Candybar	2.0 MP
Nokia 2720 fold	128 × 160 16-bit (65536) Color	2009	P	GSM	BB5.0	Clamshell	1.3 MP
Nokia 2730 classic	240 × 320 262k Color	2009	P	GSM (W-CDMA, EDGE, GPRS)	BB5.0 SL3	Candybar	2.0 MP
Nokia 2760	128 × 160 16-bit (65,536) Color	2007	D	GSM	DCT4+	Clamshell	VGA 0.3 MP
Nokia 2660	128 × 160 16-bit (65,536) Color	2009	P	EGSM	-	Flip	None
Nokia 2855i	128 × 160 262k Color (internal)/Monochromatic (external)	2005?	D	CDMA2000 1x/AMPS	-	Clamshell	None

Nokia 3000 series – Expression series

The Nokia 3000 series are mostly mid-range phones targeted towards the youth market. Some of the models in this series are targeted towards young male users, in contrast with the more unisex business-oriented 6000 series and the more feminine fashion-oriented 7000 series. Feature wise, the 3000 series slot between the 2000 and 6000 series.

Phone model	Screen type	Released	S.	Technology	Generation	Form factor	Camera
Nokia 3100	128 × 128 12-bit (4096) Color	2003	D	GSM	DCT4	Candybar	
Nokia 3105	128 × 128 12-bit (4096) Color	2003	D	CDMA2000 1x	DCT4	Candybar	
Nokia 3108	128 × 128 12-bit (4096) Color	2003	D	GSM	DCT4	Candybar	
Nokia 3109 classic	128 × 160 18-bit (262,144)	2007	D	GSM/EDGE/HSCSD	BB5.0	Candybar	
Nokia 3110	Monochrome	1997	D	GSM	DCT2	Candybar	
Nokia 3110 classic	128 × 160 18-bit (262,144)	2007	D	GSM/EDGE/HSCSD	BB5.0	Candybar	1.3M 8x Zoom
Nokia 3120	128 × 128 12-bit (4096) Color	2004	D	GSM	DCT4	Candybar	
Nokia 3125	128 × 128 12-bit (4096) Color	2004	D	CDMA2000 1x	DCT4	Candybar	
Nokia 3120 classic	240 × 320 (16M) Color	2008	D	WCDMA/GSM	-	Candybar	2M + LED Flash
Nokia 3128	128 × 160 18 bit (65,536) Color 96 × 64 16 bit (4,096) Color (external)	2004	D	GSM	DCT4	Clamshell - China market only	
Nokia 3152	128 × 160 18-bit (262,144) Color 96 × 65 Monochrome (external)	2005	D	CDMA2000 1x	DCT4	Clamshell	
Nokia 3155	128 × 160 18-bit (262,144) Color 96 × 65 Monochrome (external)	2005	D	CDMA2000 1x	DCT4	Clamshell	
Nokia 3200	128 × 128 12-bit (4096) Color	2003	D	GSM	DCT4	Candybar	
Nokia 3205	128 × 128 12-bit (4096) Color	2004	D	CDMA2000 1x/AMPS	DCT4	Candybar	
Nokia 3208 classic	240 × 320 18-bit (262,144) Color	2009	P	GSM	DCT4	Candybar	2.0 megapixels
Nokia 3210	84 × 48 Monochrome	1999	D	GSM	DCT3	Candybar	
Nokia 3220	128 × 128 16-bit (65,536) Color	2004	D	GSM	DCT4	Candybar	VGA 0.3 MP
Nokia 3230	176 × 208 16-bit (65,536) Color	2005	D	GSM	DCT4	Candybar	1.3 megapixels
Nokia 3250	176 × 208 18-bit (262,144) Color	2005	D	GSM	BB5.0	Candybar/ Twist	
Nokia 3280	84 × 48 Monochrome	2001	D	CDMA/AMPS	DCT3	Candybar	
Nokia 3285	84 × 48 Monochrome	2001	D	CDMA/AMPS	DCT3	Candybar	

Nokia 3300	128 × 128 12-bit (4096) Color	2003	D	GSM	DCT4	Candybar	
Nokia 3310	84 × 48 Monochrome	2000	D	GSM	DCT3	Candybar	
Nokia 3315	84 × 48 Monochrome	2002	D	GSM	DCT3	Candybar	
Nokia 3320	84 × 48 Monochrome	2001	D	TDMA/AMPS	DCT3	Candybar	
Nokia 3330	84 × 48 Monochrome	2001	D	GSM	DCT3	Candybar	
Nokia 3350	96 × 65 Monochrome	2001	D	GSM	DCT3	Candybar	
Nokia 3360	84 × 48 Monochrome	2001	D	TDMA/AMPS	DCT3	Candybar	
Nokia 3390	84 × 48 Monochrome	2000	D	GSM	DCT3	Candybar	
Nokia 3395	84 × 48 Monochrome	2001	D	GSM	DCT3	Candybar	
Nokia 3410	96 × 65 Monochrome	2002	D	GSM	DCT3	Candybar	
Nokia 3500 classic	128 × 160 18-bit (262,144) Color	2007	D	GSM/EDGE	BB5.0	Candybar	2.0M 8X Zoom
Nokia 3510	96 × 65 Monochrome	2002	D	GSM	DCT4	Candybar	
Nokia 3510i	96 × 65 12-bit (4096) Color	2002	D	GSM	DCT4	Candybar	
Nokia 3520	96 × 65 12-bit (4096) Color	2003	D	TDMA/AMPS	DCT4	Candybar	
Nokia 3530	96 × 65 12-bit (4096) Color	2003	D	GSM	DCT4	Candybar	
Nokia 3560	96 × 65 12-bit (4096) Color	2003	D	TDMA/AMPS	DCT4	Candybar	
Nokia 3555	128 × 160 18-bit (262,144) Color	2008	P	GSM/WCDMA EGPRS HSCSD	BB5.0	Clamshell	VGA 0.3 MP
Nokia 3570	96 × 65 Monochrome	2003	D	CDMA2000 1x	DCT4	Candybar	
Nokia 3585	96 × 65 Monochrome	2002	D	CDMA2000 1x/AMPS	DCT4	Candybar	
Nokia 3585i	96 × 65 Monochrome	2003	D	CDMA2000 1x/AMPS	DCT4	Candybar	
Nokia 3586i	96 × 65 12-bit (4096) Color	2003	D	CDMA2000 1x/AMPS	DCT4	Candybar	
Nokia 3587i	96 × 65 12-bit (4096) Color	2003	D	CDMA2000 1x/AMPS	DCT4	Candybar	
Nokia 3588i	96 × 65 12-bit (4096) Color	2003	D	CDMA2000 1x/AMPS	DCT4	Candybar	
Nokia 3589i	96 × 65 12-bit (4096) Color	2003	D	CDMA2000 1x/AMPS	DCT4	Candybar	
Nokia 3590	96 × 65 Monochrome	2002	D	GSM	DCT4	Candybar	
Nokia 3595	96 × 65 12-bit (4096) Color	2003	D	GSM	DCT4	Candybar	
Nokia 3600	176 × 208 12-bit (4096) Color	2002	D	GSM	DCT4	Candybar	
Nokia 3600 slide	240 × 320 24-bit (16 million) Color	2008	D	GSM	BB5.0	Slide	3.2 megapixels
Nokia 3606	176 × 220 24-bit (16 million) Color	2008	P	CDMA	BB5.0	Clamshell	1.3 megapixels
Nokia 3610	96 × 65 Monochrome	2002	D	GSM	DCT3	Candybar	
Nokia 3620	176 × 208 16-bit (65,536) Color	2003	D	GSM	DCT4	Candybar	
Nokia 3650	176 × 208 12-bit (4096) Color	2003	D	GSM	DCT4	Candybar/ Circular keypad	VGA 0.3 MP
Nokia 3660	176 × 208 16-bit (65,536) Color	2003	D	GSM	DCT4	Candybar	VGA 0.3 MP

Nokia 3810	Monochrome	1997	D	GSM	DCT3	Candybar	

The Nokia 4000 series was skipped as a sign of deference from Nokia towards East Asian customers.[2]

Nokia 5000 series – Active series

The Nokia 5000 series are similar in features to the 3000 series, but often contain more features towards active individuals. Many of the 5000 series phones feature a rugged construction or contain extra features for music playback.

Phone model	Screen type	Released	S.	Technology	Generation	Form factor	Camera
Nokia 5000	240 × 320 (65,000) Color	2008	D	GSM		Candybar	1.3 MP
Nokia 5030 XpressRadio	128 × 160 16 bit (65,536) Color	2009	P	GSM		Candybar	None
Nokia 5070	128 × 160 16-bit (65,536) Color	2007	D	GSM	DCT4	Candybar	VGA(0.3 MP)
Nokia 5100	128 × 128 12-bit (4096) Color	2003	D	GSM	DCT4	Candybar	None
Nokia 5110	84 × 48 Monochrome	1998	D	GSM	DCT3	Candybar	None
Nokia 5120	84 × 48 Monochrome	1998	D	TDMA/AMPS	DCT3	Candybar	None
Nokia 5125	84 × 48 Monochrome	1998	D	TDMA/AMPS	DCT3	Candybar	None
Nokia 5130	84 × 48 Monochrome	1998	D	GSM	DCT3	Candybar	None
Nokia 5130 XpressMusic	240 × 320 (256,000) Color	2008	D	GSM		Candybar	2.0 MP
Nokia 5140	128 × 128 12-bit (4096) Color	2004	D	GSM	DCT4	Candybar	None
Nokia 5140i	128 × 128 16-bit (65,536) Color	2005	D	GSM	DCT4	Candybar	VGA (0.3 MP)
Nokia 5160	84 × 48 Monochrome	1998	D	TDMA/AMPS	DCT3	Candybar	None
Nokia 5165	84 × 48 Monochrome	2000	D	TDMA/AMPS	DCT3	Candybar	None
Nokia 5170	84 × 48 Monochrome	1999	D	CDMA	DCT3	Candybar	None
Nokia 5170i	84 × 48 Monochrome	1999	D	CDMA	DCT3	Candybar	None
Nokia 5180iP	84 × 48 Monochrome	2002	D	CDMA/AMPS	DCT3	Candybar	None
Nokia 5185i	84 × 48 Monochrome	2002	D	CDMA/AMPS	DCT3	Candybar	None
Nokia 5190	84 × 48 Monochrome	1998	D	GSM	DCT3	Candybar	None

Nokia 5200 XpressMusic	128 × 160 18-bit (262,144) Color	2006	D	GSM	BB5.0	Slide	VGA (0.3 MP)
Nokia 5210	84 × 48 Monochrome	2002	D	GSM	DCT3	Candybar	None
Nokia 5220 XpressMusic	240 × 320 (262K) Color	2008	D	GSM		Candybar	2.0 MP
Nokia 5230	360 × 640 24-bit (16M) Color	2009	P	UMTS/GSM/EDGE/HSPDA/	BB5.0	Candybar	2.0 MP
Nokia 5233	360 × 640 24-bit (16M) Color	2010	P	GSM/EDGE	BB5.0	Candybar	2.0 MP
Nokia 5250	640 × 360 24-bit (16M) Color	2010	P	UMTS/GSM/EDGE/HSPDA/	BB5.0	Candybar	2.0 MP
Nokia 5300 XpressMusic	320 × 240 18-bit (262,144) Color	2006	D	GSM	BB5.0	Slide	1.3 MP
Nokia 5310 XpressMusic	240 × 320 24 bit (16,777,216) Color	2007	D	GSM	BB5.0	Candybar	2.0 MP
Nokia 5320 XpressMusic	240 × 320 (16M) Color	2008	D	UMTS		Candybar	2.0 MP + CIF
Nokia 5330 Mobile TV Edition	240 × 320 24 bit (16,777,216) Color	2009	P	UMTS	BB5.0	Slide	3.2 MP
Nokia 5500	208 × 208 18-bit (262,144) Color	2006	D	GSM	BB5.0	Candybar	2.0 MP
Nokia 5510	84 × 48 Monochrome	2001	D	GSM	DCT3	Candybar	None
Nokia 5530 XpressMusic	360 × 640 24-bit (16M) Color	2009	P	GSM/EDGE/WLAN	BB5.0	Candybar	3.2 MP + VGA
Nokia 5610 XpressMusic	240 × 320 24 bit (16,777,216) Color	2007	D	UMTS/GSM/EDGE	BB5.0	Slide	3.2 MP
Nokia 5630 XpressMusic	240 × 320 24 bit (16,777,216) Color	2009	P	UMTS	BB5.0	Candybar	3.2 MP
Nokia 5700 XpressMusic	320 × 240 24-bit (16M) Color	2007	D	UMTS/GSM/EDGE	BB5.0	Candybar/ Twist	2.0 MP
Nokia 5730 XpressMusic	240 × 320 24 bit (16,777,216) Color	2009	P	UMTS	BB5.0	Bar w/ Slide-out QWERTY keyboard	3.2 MP + VGA
Nokia 5800 XpressMusic (AKA Nokia 5800 Navigation Edition)	360 × 640 24-bit (16M) Color	2008	D	UMTS/GSM/EDGE/HSDPA	BB5.0	Candybar	3.2 MP + VGA

The 5210 features rubber Xpress-On shells, WAP over CSD and a built in thermometer. The thermometer is actually the internal temperature of the phone's battery, this feature is also present on other phones that have "netmonitor" enabled. The 5210 is nicknamed a "builder's phone" because of its rubber splash/impact proof casing. Its successor is the 5100 and after that, the 5140 and 5140i.

The 5510 was Nokia's first phone with a built in MP3 player, and it had 64 megabytes of memory for storing MP3s. It also had a full QWERTY keyboard and an 84 × 48 monochrome display. This phone did not sell very well even though it was advertised on television, possibly because it was too expensive and too big. Its replacement is the Nokia 3300.

The 5330 XpressMusic was discontinued, but replaced by the Nokia X3-00. The X3-00 features some styling cues taken from the 5330.

Nokia 6000 series – Classic Business series

The Nokia 6000 series is Nokia's largest family of phones. It consists mostly of mid-range to high-end phones containing a high amount of features. The 6000 series is notable for their conservative, unisex designs, which make them popular among business users.

Phone model	Screen type	Released	S.	Technology	Generation	Form factor	Camera
Nokia 6010	96 × 65 12-bit (4096) Color	2004	D	GSM	DCT4	Candybar	None
Nokia 6011i	96 × 65 12-bit (4096) Color	2004	D	CDMA2000 1x	DCT4	Candybar	None
Nokia 6012	96 × 65 12-bit (4096) Color	2004	D	CDMA2000 1x/AMPS	DCT4	Candybar	None
Nokia 6015	96 × 65 12-bit (4096) Color	2004	D	CDMA2000 1x/AMPS	DCT4	Candybar	None
Nokia 6015i	96 × 65 12-bit (4096) Color	2004	D	CDMA2000 1x/AMPS	DCT4	Candybar	None
Nokia 6016i	96 × 65 12-bit (4096) Color	2004	D	CDMA2000 1x/AMPS	DCT4	Candybar	None
Nokia 6019i	96 × 65 12-bit (4096) Color	2004	D	CDMA2000 1x/AMPS	DCT4	Candybar	None
Nokia 6020	128 × 128 16-bit (65,536) Color	2005	D	GSM	DCT4	Candybar	0.3
Nokia 6021	128 × 128 16-bit (65,536) Color	2005	D	GSM	DCT4	Candybar	None
Nokia 6030	128 × 128 16-bit (65,536) Color	2005	D	GSM	DCT4	Candybar	None
Nokia 6060	128 × 160 16-bit (65,536) Color	2005	D	GSM	DCT4	Clamshell	None
Nokia 6070	128 × 160 16-bit (65,536) Color	2006	D	GSM	DCT4	Candybar	0.3
Nokia 6085	128 × 160 18-bit (262,144) Color	2006	D	GSM	BB5.0	Clamshell	0.3
Nokia 6086	128 × 160 18-bit (262,144) Color	2007	D	GSM/UMA (VoIP)	DCT4	Clamshell	0.3
Nokia 6090 (car phone)	84 × 48 Monochrome	1999	D	GSM	DCT3	Candybar	None
Nokia 6100	128 × 128 12-bit (4096) Color	2002	D	GSM	DCT4	Candybar	None
Nokia 6101	128 × 160 16-bit (65,536) Color	2005	D	GSM	DCT4	Clamshell	0.3
Nokia 6102	128 × 160 16-bit (65,536) Color	2005	D	GSM	DCT4	Clamshell	0.3

Nokia 6102i	128 × 160 16-bit (65,536) Color	2006	D	GSM	DCT4	Clamshell	0.3
Nokia 6103	128 × 160 16-bit (65,536) Color	2006	D	GSM	DCT4	Clamshell	0.3
Nokia 6108	128 × 128 12-bit (4096) Color	2003	D	GSM	DCT4	Candybar	None
Nokia 6110	84 × 48 Monochrome	1997	D	GSM	DCT3	Candybar	None
Nokia 6110 Navigator	240 × 320 24-bit (16,777,216) Color	2007	D	HSDPA/GSM/EDGE	BB5.0	Slide	2.0
Nokia 6111	128 × 160 18-bit (262,144) Color	2005	D	GSM	DCT4	Slide	1.0
Nokia 6120	84 × 48 Monochrome	1998	D	TDMA/AMPS	DCT3	Candybar	None
Nokia 6120 Classic	240 × 320 24-bit (16,777,216) Color	2007	D	HSDPA/GSM/EDGE	S60 3rd Edition	Candybar	2.0
Nokia 6121 Classic	240 × 320 24-bit (16,777,216) Color	2007	D	HSDPA/GSM/EDGE	BB5.0	Candybar	2.0, 4X digital zoom
Nokia 6122 Classic	240 × 320 24-bit (16,777,216) Color	2008	P	GSM/EDGE	BB5.0	Candybar	2.0
Nokia 6125	128 × 160 18-bit (262,144) Color	2006	D	GSM	BB5.0	Clamshell	None
Nokia 6126	320 × 240 24-bit (16,777,216) Color	2006	D	GSM	BB5.0	Clamshell	1.3
Nokia 6130	84 × 48 Monochrome	1997	D	GSM	DCT3	Candybar	None
Nokia 6131	320 × 240 24-bit (16,777,216) Color	2006	D	GSM	BB5.0	Clamshell	1.3
Nokia 6133	320 × 240 24-bit (16,777,216) Color	2006	D	GSM	BB5.0	Clamshell	1.3
Nokia 6136	128 × 160 18-bit (262,144) Color	2006	D	GSM/UMA (VoIP)	BB5.0	Clamshell	None
Nokia 6138	84 × 48 Monochrome	1997	D	GSM	DCT3	Candybar	None
Nokia 6162	84 × 48 Monochrome	1998	D	TDMA/AMPS	DCT3	Clamshell	None
Nokia 6150	84 × 48 Monochrome	1998	D	GSM	DCT3	Candybar	None
Nokia 6151	128 × 160 (262,144) Color	2006	D	GSM/WCDMA	BB5.0	Candybar	1.3
Nokia 6155	128 × 160 18-bit (262,144) Color 96 × 65 16-bit (65,536) Color (external)	2005	D	CDMA2000 1x/AMPS	DCT4	Clamshell	None
Nokia 6155i	128 × 160 18-bit (262,144) Color 96 × 65 16-bit (65,536) Color (external)	2005	D	CDMA2000 1x/AMPS	DCT4	Clamshell	None
Nokia 6160	84 × 48 Monochrome	1998	D	TDMA/AMPS	DCT3	Candybar	None

Nokia 6165i	128 × 160 18-bit (262,144) Color 96 × 65 16-bit (65,536) Color (external)	2005	D	CDMA2000 1x/AMPS	DCT4	Clamshell	None
Nokia 6170	128 × 160 16-bit (65,536) Color	2004	D	GSM	DCT4	Clamshell	None
Nokia 6185	Monochrome	1999	D	CDMA/AMPS	DCT3	Candybar	None
Nokia 6190	84 × 48 Monochrome	1997	D	GSM	DCT3	Candybar	None
Nokia 6200	128 × 128 12-bit (4096) Color	2003	D	GSM	DCT4	Candybar	None
Nokia 6210	96 × 65 Monochrome	2000	D	GSM	DCT3	Candybar	None
Nokia 6210 Navigator	240 × 320 24-bit (16,777,216) Color	2008	P	HSDPA/GSM/EDGE	BB5.0	Slide	3.2
Nokia 6215i	128 × 128 18-bit (262,144) Color Organic light-emitting diode (OLED)	2006	D	CDMA	DCT4	Clamshell	None
Nokia 6220	128 × 128 12-bit (4096) Color	2003	D	GSM	DCT4	Candybar	Yes
Nokia 6220 classic	240 × 320 24-bit (16,777,216) Color 2.2	2008	P	HSDPA/GSM , WCDMA/EDGE	BB5.0	Candybar	5.0, Carl Zeiss, Zenon Flash
Nokia 6223						Candybar	None
Nokia 6225	128 × 128 12-bit (4096) Color	2004	D	CDMA2000 1x/AMPS	DCT4	Candybar	0.3
Nokia 6230	128 × 128 16-bit (65,536) Color	2004	D	GSM	DCT4	Candybar	0.3
Nokia 6230i	208 × 208 16-bit (65,536) Color	2005	D	GSM	DCT4	Candybar	1.3
Nokia 6233	320 × 240 18-bit (262,144) Color	2006	D	UMTS/GSM	BB5.0	Candybar	2.0
Nokia 6234	320 × 240 18-bit (262,144) Color	2006	D	UMTS/GSM	BB5.0	Candybar	2.0
Nokia 6235/6236	128 × 128 16-bit (65,536) Color	2005	D	CDMA2000 1x	DCT4	Candybar	0.3
Nokia 6235i/6236i	128 × 128 16-bit (65,536) Color	2005	D	CDMA2000 1x/AMPS	DCT4	Candybar	0.3
Nokia 6250	96 × 65 Monochrome	2000	D	GSM	DCT3	Candybar	None
Nokia 6255i	128 × 160 16-bit (65,536) Color	2004	D	CDMA2000 1x/AMPS	DCT4	Clamshell	0.3
Nokia 6256i	128 × 160 16-bit (65,536) Color	2004	D	CDMA2000 1x/AMPS	DCT4	Clamshell	0.3
Nokia 6260	176 × 208 16-bit (65,536) Color	2004	D	GSM	DCT4	Clamshell	0.3
Nokia 6260 Slide	320 × 480 24-bit (16.7M) Color	2008	P	GSM/WCDMA/HSDPA/HSUPA/VoIP	BB5	Slide	5.0, Carl Zeiss optics, Flash

Nokia 6263	240 × 320 24-bit (16.7M) Color	2008	P	GSM/WCDMA EGPRS HSCSD	BB5.0	Clamshell	1.0
Nokia 6265	320 × 240 18-bit (262,144) Color	2005	D	CDMA2000 1x	DCT4	Slide	2.0
Nokia 6267	320 × 240 24-bit (16,777,216) Color	2007	D	UMTS/GSM	BB5.0	Clamshell	2.0
Nokia 6270	320 × 240 18-bit (262,144) Color	2005	D	GSM	BB5.0	Slide	2.0
Nokia 6275	320 × 240 18-bit (262,144) Color	2006	D	CDMA2000 1x	BB5.0	Candybar	2.0
Nokia 6280	320 × 240 18-bit (262,144) Color	2005	D	UMTS/GSM	BB5.0	Slide	2.0
Nokia 6288	320 × 240 18-bit (262,144) Color	2006	D	UMTS/GSM	BB5.0	Slide	2.0
Nokia 6290	240 × 320 24-bit (16,777,216) Color 128 × 160 16-bit (262,144) Color (external)	2006	D	UMTS/GSM	BB5.0	Clamshell	2.0
Nokia 6300	240 × 320 24 bit (16,777,216) Color	2006	D	GSM	BB5.0	Candybar	2.0
Nokia 6303 Classic	320 × 240 32 bit (16.7M) colour	2009	D	GSM	BB5.92	Classic	3.2, autofocus and flash
Nokia 6300i	240 × 320 24 bit (16,777,216) Color	2008	D	GSM and VoIP	BB5.0	Candybar	2.0
Nokia 6301		2007	D	GSM/UMA	BB5.0	Candybar	None
Nokia 6310	96 × 65 Monochrome	2001	D	GSM	DCT4	Candybar	None
Nokia 6310i	96 × 65 Monochrome	2002	D	GSM	DCT4	Candybar	None
Nokia 6315i	128 × 160 18-bit (262,144) Color	2006	D	CDMA	DCT4	Clamshell	None
Nokia 6340	96 × 65 Monochrome	2003	D	GAIT	DCT4	Candybar	None
Nokia 6340i	96 × 65 Monochrome	2002	D	GAIT	DCT4	Candybar	None
Nokia 6350	240 × 320 16-bit	2009	D	HSDPA	BB5.0	Clamshell	2.0
Nokia 6360	96 × 65 Monochrome	2001	D	TDMA	DCT4	Candybar	None
Nokia 6370	96 × 65 Monochrome	2002	D	CDMA2000 1x	DCT4	Candybar	None
Nokia 6385	96 × 65 Monochrome	2002	D	CDMA2000 1x/AMPS	DCT4	Candybar	None
Nokia 6500	96 × 65 Monochrome	2002	D	GSM	DCT4	Candybar (slide cover)	None
Nokia 6500 classic	240 × 320 24-bit (16.7 million) Color	2007	D	GSM/UMTS	BB5.0	Candybar	2.0
Nokia 6500 slide	240 × 320 24-bit (16.7 million) Color	2007	D	GSM/UMTS	BB5.0	Slide	3.2, Carl Zeiss optics, Auto-Focus
Nokia 6510	96 × 65 Monochrome	2002	D	GSM	DCT4	Candybar	None
Nokia 6555	240 × 320 24-bit (16.7 million) Color	2007	D	GSM/UMTS	BB5.0	Clamshell	1.3

Nokia 6560	128 × 128 12-bit (4096) Color	2003	D	TDMA/AMPS	DCT4	Candybar	None
Nokia 6585	128 × 128 12-bit (4096) Color	2003	D	CDMA2000 1x/AMPS	DCT4	Candybar	None
Nokia 6590	96 × 65 Monochrome	2002	D	GSM	DCT4	Candybar	None
Nokia 6590i	96 × 65 Monochrome	2003	D	GSM	DCT4	Candybar	None
Nokia 6600	176 × 208 16-bit (65,536) Color	2003	D	GSM	DCT4	Candybar	0.3
Nokia 6600 fold	240 × 320 24-bit (16 million) OLED Color	2008	D	UMTS/GSM	BB5.0	Clamshell	2.0
Nokia 6600 slide	240 × 320 24-bit (16 million) Color	2008	D	UMTS/GSM	BB5.0	Slide	3.2
Nokia 6600i slide	240 × 320 24-bit (16 m) Color	2009	P	UMTS/GSM	BB5.0	Slide	5.0
Nokia 6610	128 × 128 12-bit (4,096) Color	2002	D	GSM	DCT4	Candybar	None
Nokia 6610i	128 × 128 12-bit (4,096) Color	2004	D	GSM	DCT4	Candybar, CIF	0.1
Nokia 6620	176 × 208 16-bit (65,536) Color	2004	D	GSM	DCT4	Candybar	None
Nokia 6630	176 × 208 16-bit (65,536) Color	2004	D	UMTS/GSM	BB5.0	Candybar	1.3
Nokia 6638	176 × 208 16-bit (65,536) Color	2004	D	CDMA2000 1x	BB5.0	Candybar	None
Nokia 6650	128 × 160 12-bit (4096) Color	2003	D	UMTS/GSM	DCT4	Candybar	None
Nokia 6650 fold	240 × 320 18-bit (262,144) Color	2008	P	HSDPA/GSM	S60 3rd Edition, Feature Pack 2 (i.e. version 3.2)	Clamshell	2.0
Nokia 6651	128 × 160 12-bit (4096) Color	2004	D	UMTS/GSM	DCT4	Candybar	None
Nokia 6670	176 × 208 16-bit (65,536) Color	2004	D	GSM	DCT4	Candybar	1.0
Nokia 6700 classic	240 × 320 24-bit (16,777,216) Color	2009	P	HSDPA/GSM	BB5.0	Candybar	5.0
Nokia 6700 slide	240 × 320 24-bit (16,777,216) Color	2010	P	HSDPA/GSM	S60 3rd Edition FP2	Slide	5.0
Nokia 6710 Navigator	240 × 320 24-bit (16,777,216) Color	2009	P	HSDPA/GSM/EDGE	S60 3rd Edition FP2	Slide	5.0, Autofocus
Nokia 6720	240 × 320 24 bit (16,777,216) Color	2009	P	HSDPA/GSM	S60 3rd Edition FP2	Candybar	5.0, Autofocus
Nokia 6730	240 × 320 24-bit (16,777,216) Color	2009	P	GSM/EDGE/UMTS	S60 3rd Edition FP2	Candybar	3.2
Nokia 6750 Mural	240 × 320 24-bit (16,777,216) Color	2009	P	GSM/EDGE/UMTS/HSCSD/HSDPA	Series 40 6th Edition	Clamshell	2.0
Nokia 6760 slide	320 × 240 24-bit (16,7 m) Color	2009	P	HSDPA/GSM	S60 3rd Edition FP2	QWERTY BOARD	3.2

Nokia 6680	176 × 208 18-bit (262,144) Color	2005	D	UMTS/GSM	BB5.0	Candybar	1.3
Nokia 6681	176 × 208 18-bit (262,144) Color	2005	D	GSM	BB5.0	Candybar	1.3
Nokia 6682	176 × 208 18-bit (262,144) Color	2005	D	GSM	BB5.0	Candybar	1.3
Nokia 6800	128 × 128 12-bit (4096) Color	2003	D	GSM	DCT4	Candybar (flip keyboard)	None
Nokia 6810	128 × 128 12-bit (4096) Color	2004	D	GSM	DCT4	Candybar (flip keyboard)	None
Nokia 6820	128 × 128 12-bit (4096) Color	2004	D	GSM	DCT4	Candybar (flip keyboard)	None
Nokia 6822	128 × 128 16-bit (65,536) Color	2005	D	GSM	DCT4	Candybar (flip keyboard)	None

Nokia 6136 UMA is the first mobile phone to include Unlicenced Mobile Access. Nokia 6131 NFC is the first mobile phone to include Near Field Communication.

Nokia 7000 series – Fashion and Experimental series

The Nokia 7000 series is a family of Nokia phones with two uses. Most phones in the 7000 series are targeted towards fashion-conscious users, particularly towards women. Some phones in this family also test new features. The 7000 series are considered to be a more consumer-oriented family of phones when contrasted to the business-oriented 6000 series.

Phone model	Screen resolution	Screen type	Screen bit colour	Released	S.	Technology	Generation	Form factor	Camera (Mpix.)
Nokia 7070 Prism	128 × 160	Color		2008	D	GSM		Clamshell	None
Nokia 7100 Supernova	240 × 320	Color		2008	D	GSM		Slide	1.3
Nokia 7110	96 × 65	Monochrome	1	1999	D	GSM	DCT3	Candybar (slide cover)	None
Nokia 7160	96 × 65	Monochrome	1	2000	D	TDMA/AMPS	DCT3	Candybar (slide cover)	None
Nokia 7190	96 × 65	Monochrome	1	2000	D	GSM	DCT3	Candybar (slide cover)	None
Nokia 7200	128 × 128	Color	16	2004	D	GSM	DCT4	Clamshell	0.3
Nokia 7210	128 × 128	Color	12	2002	D	GSM	DCT4	Candybar	None
Nokia 7210 Supernova	240 x 320	Color	18	2008	D	GSM	DCT4	Candybar	2.0
Nokia 7230	240 x 320	Color	18	2010	P	GSM/UMTS	DCT4	Slide	3.2
Nokia 7250	128 × 128	Color	12	2003	D	GSM	DCT4	Candybar	0.1
Nokia 7250i	128 × 128	Color	12	2003	D	GSM	DCT4	Candybar	0.1
Nokia 7260	128 × 128	Color	16	2004	D	GSM	DCT4	Candybar	None

Nokia 7270	128 × 160	Color	16	2004	D	GSM	DCT4	Clamshell	0.3
Nokia 7280	104 × 208	Color	16	2004	D	GSM	DCT4	Lipstick / Slide	None
Nokia 7310 Supernova	240 × 320	Color	18	2008	D	GSM	DCT4	Candybar	2.0
Nokia 7360	128 × 160	Color	16	2006	D	GSM	DCT4	Candybar	0.3
Nokia 7370	240 × 320	Color	18	2006	D	GSM	BB5.0	Swivel	1.3
Nokia 7373	240 × 320	Color	18	2006	D	GSM/UMTS	BB5.0	Swivel	2.0
Nokia 7380	104 × 208	Color	16	2006	D	GSM	DCT4	Lipstick	2.0
Nokia 7390	240 × 320	Color	24	2006	D	GSM/UMTS	BB5.0	Clamshell	3.0
Nokia 7500 Prism	240 × 320	Color		2007	D	GSM		Candybar	2.0
Nokia 7510 Supernova	240 × 320	Color	24	2009	D	GSM/UMA	BB5.0	Clamshell	2.0
Nokia 7600	128 × 160	Color	16	2003	D	GSM/UMTS	DCT4	Square Candybar	0.3
Nokia 7610 Supernova	240 × 320	Color	16	2008	D	GSM	DCT4	Slide	3.2
Nokia 7650	176 × 208	Color	12	2002	D	GSM	DCT4	Slide	0.3
Nokia 7700	640 × 320	Color	16	Cancelled		GSM	DCT4 (APE)	Candybar	0.3
Nokia 7710	640 × 320	Color	16	2005	D	GSM	DCT4 (APE)	Candybar	1.0
Nokia 7900 Prism	240 × 320	Color	16	2007	D	GSM		Candybar	2.0
Nokia 7900 Crystal Prism	240 × 320	Color	24	2008	D	GSM		Candybar	2.0

The 7110 was the first Nokia phone with a WAP browser. WAP was significantly hyped up during the 1998–2000 Internet boom. However WAP did not meet these expectations and uptake was limited. Another industry first was the flap, which slid from beneath the phone with a push from the release button. Unfortunately the cover was not too durable. The 7110 was also the only phone to feature a navi-roller key.

The 7250i was a slightly improved version of the Nokia 7250. It includes XHTML and OMA Forward lock digital rights management. The phone has exactly the same design as the 7250. This phone is far more popular than the 7250 and has been made available on pre-paid packages and therefore it is very popular amongst youths in the UK and other European countries.

The 7510 Supernova was a phone exclusive to T-Mobile USA. Only some newer units of this model have Wi-Fi chips with UMA. The Wi-Fi adapter on this phone supports up to WPA2 encryption if present. This phone uses Xpress-On Covers.

The 7650 was the first Series 60 smartphone of Nokia. It was quite basic compared to new smartphones, it didn't have MMC slot, but it had a camera.

The 7610 was Nokia's first smartphone featuring a megapixel camera (1,152×864 pixels), and is targeted towards the fashion conscious individual. End-users can also use the 7610 with Nokia Lifeblog. Other pre-installed applications include the Opera and Kodak Photo Sharing. It is notable for its looks, having opposite corners rounded off. It comes with a 64 MB Reduced Size MMC. The main CPU is an ARM compatible chip (ARM4T architecture) running at 123 MHz.

The 7710's 640 × 320 screen is a touch screen.

Nokia 8000 series – Premium series

This series is characterized by ergonomics and attractiveness. The internals of the phone are similar to those in different series and so on that level offer nothing particularly different, however the physical handset itself offers a level of functionality which appeals to users who focus on ergonomics. The front slide keypad covers offered a pseudo-flip that at the time Nokia were unwilling to make. Materials used increased the cost and hence exclusivity of these handsets.

The only exception to the rule (there are many in different series) is the 82xx, 8310 which were very small and light handsets.

Phone model	Screen type	Released	S.	Technology	Generation	Form factor
Nokia 8110	Monochrome	1996	D	GSM	DCT2	Candybar (slide cover)
Nokia 8148	Monochrome	1996	D	GSM	DCT2	Candybar (slide cover)
Nokia 8210	84 × 48 Monochrome	1999	D	GSM	DCT3	Candybar
Nokia 8250	84 × 48 Monochrome	2001	D	GSM	DCT3	Candybar
Nokia 8260	84 × 48 Monochrome	2000	D	TDMA/AMPS	DCT3	Candybar
Nokia 8265	84 × 48 Monochrome	2002	D	TDMA/AMPS	DCT3	Candybar
Nokia 8265i	84 × 48 Monochrome	2002	D	TDMA/AMPS	DCT3	Candybar
Nokia 8270	84 × 48 Monochrome	2002	D	CDMA	DCT3	Candybar
Nokia 8280	84 × 48 Monochrome	2002	D	CDMA2000 1x	DCT3	Candybar
Nokia 8290	84 × 48 Monochrome	2001	D	GSM	DCT3	Candybar
Nokia 8310	84 × 48 Monochrome	2001	D	GSM	DCT4	Candybar
Nokia 8390	84 × 48 Monochrome	2002	D	GSM	DCT4	Candybar
Nokia 8600	320 × 240 24-bit (16.7M) Color	2007	D	GSM	BB5.0	Slider
Nokia 8800	208 × 208 18-bit (262,144) Color	2005	D	GSM	DCT4	Soft-slide Stainless Steel cover
Nokia 8801	208 × 208 18-bit (262,144) Color	2005	D	GSM	DCT4	Soft-slide Stainless Steel cover
Nokia 8810	Monochrome	1998	D	GSM	DCT3	Soft-slide Chrome cover
Nokia 8850	84 × 48 Monochrome	1999	D	GSM	DCT3	Candybar (slide cover)
Nokia 8855	84 × 48 Monochrome	2001	D	GSM	DCT3	Candybar (slide cover)
Nokia 8860	84 × 48 Monochrome	1999	D	TDMA/AMPS	DCT3	Candybar (slide cover)
Nokia 8890	84 × 48 Monochrome	2000	D	GSM	DCT3	Candybar (slide cover)
Nokia 8910	84 × 48 Monochrome	2002	D	GSM	DCT4	Auto-slide Titanium cover
Nokia 8910i	96 × 65 12-bit (4096) Color	2003	D	GSM	DCT4	Auto-slide Titanium cover

Nokia 9000 series – Communicator series (discontinued)

The Nokia 9000 series was reserved for the Communicator series, but the latest Communicator, the E90 Communicator, is an Eseries phone.

Phone model	Screen type	Released	S.	Technology	Generation	Form factor
Nokia 9000 Communicator	640 × 200 Monochrome	1996	D	GSM	DCT1	Clamshell
Nokia 9000i Communicator	640 × 200 Monochrome	1997	D	GSM	DCT1	Clamshell
Nokia 9110 Communicator	640 × 200 Monochrome	1998	D	GSM	DCT3	Clamshell
Nokia 9110i Communicator	640 × 200 Monochrome	2000	D	GSM	DCT3	Clamshell
Nokia 9210 Communicator	640 × 200 12-bit (4096) Color	2001	D	GSM	DCTL	Clamshell
Nokia 9210i Communicator	640 × 200 12-bit (4096) Color	2002	D	GSM	DCTL	Clamshell
Nokia 9290 Communicator	640 × 200 12-bit (4096) Color	2002	D	GSM	DCTL	Clamshell
Nokia 9300	640 × 200 16-bit (65,536) Color	2005	D	GSM	DCT4 (APE)	Clamshell
Nokia 9300i	640 × 200 16-bit (65,536) Color	2006	D	GSM	DCT4 (APE)	Clamshell
Nokia 9500 Communicator	640 × 200 16-bit (65,536) Color	2004	D	GSM	DCT4 (APE)	Clamshell

Modern series (C/E/N/X/Asha/Lumia)

Nokia Cseries

The Nokia Cseries is an affordable series optimized for social networking and sharing.

Phone model	Screen type	Released	S.	Technology	Platform	Generation	Form factor	Camera
Nokia C1-00(01 & 02)	128 × 160 pixels (65K)	2010 Q1	U	GSM	Series 30 (C1-00) Series 40 6th Edition (C1-01 & C1-02)	XGOLD 213	Candybar	VGA (0.3 MP) (C1-01)
Nokia C2	128 × 160 pixels (65K)		U	GSM/UMTS	Series 40 6th Edition	XGOLD 213	Candybar	VGA (0.3MP)
Nokia C3-00	320 × 240 pixels (256K) Color TFT	2010 Q2	P	GSM/UMTS WLAN	Series 40 6th Edition	BB5.0	QWERTY Candybar	2.0 MP
Nokia C3-01 (Touch and Type)	240 × 320 (256K) TFT Color	2010 Q4	U	GSM EDGE, UMTS, WLAN	Series 40 6th Edition feature pack 1	BB5.0	Candybar	5.0 MP
Nokia C3-01 Gold Edition (Touch and Type)	240 × 320 (256K) TFT Color	2010 Q4	U	GSM EDGE, UMTS, WLAN	Series 40 6th Edition feature pack 1	BB5.0	Candybar	5.0 MP
Nokia C5-00	240 × 320 pixels (16.7M) Color TFT	2010 Q2	P	GSM/UMTS	S60 3rd Edition FP2	BB5.0	Candybar	3.2 MP
Nokia C5-03	640 × 360 pixels (16.7M) transmissive	2010 Q4	P	GSM EGPRS UMTS WLAN WCDMA/HSDPA EGSM	S60 5th Edition	tbc	Touchscreen candybar	5.0 MP
Nokia C6-00	640 × 360 pixels (16.7M)	2010 Q2	P	GSM EGPRS UMTS WLAN WCDMA/HSDPA EGSM	S60 5th Edition	BB5.0	QWERTY Slider	5.0 MP

Nokia C6-01	640 × 360 pixels (16.7M)	2010 Q4	P	GSM EGPRS UMTS WLAN WCDMA/HSDPA EGSM	Symbian^3	tbc	Touchscreen candybar	8.0 MP (720p HD)
Nokia C7-00	640 × 360 pixels (16.7M)	2010 Q4	P	GSM EGPRS UMTS WLAN WCDMA/HSDPA EGSM	Symbian^3	tbc	Touchscreen Monoblock	8.0 MP (720p HD)

C1-00 and C2-00 are Dual-SIM phones, but for the Nokia C1-00, both SIM cards cannot be utilized at the same time.

Nokia Eseries

The Nokia Eseries is an enterprise-class series and includes business-optimized smartphones.

Phone model	Screen type	Released	S.	Technology	Platform	Generation	Form factor	Camera	Images
Nokia E50	240 × 320 18-bit (262,144) Color	2006	D	GSM/EDGE	S60 3rd Edition	BB5.0	Candybar	1.3 megapixels	
Nokia E51	240 × 320 24-bit (16.7 million) Color	2007	D	GSM/EDGE/HSDPA/3G/WLAN	S60 3rd Edition FP1	BB5.0	Candybar	2 megapixels	
Nokia E51 (No Camera)	240 × 320 24-bit (16.7 million) Color	2008	D	GSM/EDGE/HSDPA/3G/WLAN	S60 3rd Edition FP1	BB5.0	Candybar	-	
Nokia E52	320 × 240 24-bit (16.7 million) Color	2009	P	GSM/EDGE/HSDPA/3G/WLAN	S60 3rd Edition FP2	BB5.0	Candybar	3.2 megapixels	
Nokia E55	320 × 240 24-bit (16.7 million) Color	2009	P	GSM/EDGE/HSDPA/3G/WLAN	S60 3rd Edition FP2	BB5.0	Candybar	3.2 megapixels	
Nokia E60	352 × 416 24-bit (16.7 million) Color	2006	D	GSM/UMTS/WLAN	S60 3rd Edition	BB5.0	Candybar	None	

Nokia E61	320 × 240 24-bit (16.7 million) Color	2006	D	GSM/UMTS/WLAN	S60 3rd Edition	BB5.0	Candybar	None	
Nokia E61i	320 × 240 24-bit (16.7 million) Color	2007	D	GSM/EDGE/UMTS/WLAN	S60 3rd Edition	BB5.0	Candybar	2 megapixels	
Nokia E62	320 × 240 24-bit (16.7 million) Color	2006	D	GSM/EDGE	S60 3rd Edition	BB5.0	QWERTY Candybar	None	
Nokia E63	320 × 240 24-bit (16.7 million) Color	2008	P	GSM/EDGE/UMTS/WLAN	S60 3rd Edition FP2	BB5.0	QWERTY Candybar	2 megapixels	
Nokia E65	320 × 240 24-bit (16.7 million) Color	2007	D	GSM/EDGE/UMTS/WLAN	S60 3rd Edition	BB5.0	Slider	2 megapixels	
Nokia E66	240 × 320 24-bit (16.7 million) Color	2008	P	GSM/EDGE/UMTS/WLAN	S60 3rd Edition FP1	BB5.0	Slider	3.15 megapixels	
Nokia E6	640 × 480 24-bit (16.7 million) Color	2011	U	GSM/EDGE/UMTS/WLAN	Symbian^3 'Anna'(PR 2.0)	tbc	QWERTY candybar	8 megapixels	
Nokia E70	352 × 416 24-bit (16.7 million) Color	2006	D	GSM/WLAN/UMTS (Europe/Asia)	S60 3rd Edition	BB5.0	Candybar (flip keyboard)	2 megapixels	
Nokia E71	320 × 240 24-bit (16.7 million) Color	2008	P	GSM/EDGE/UMTS/WLAN	S60 3rd Edition FP1	BB5.0	Candybar	3.2 megapixels	

Nokia E72	320 × 240 24-bit (16.7 million) Color	2009	P	GSM/EDGE/HSDPA/HSUPA/WLAN	S60 3rd Edition FP2	BB5.0	Qwerty Candybar	5.0 megapixels	
Nokia E73 Mode	320 × 240 24-bit (16.7 million) Color	2010	P	GSM/UMTS, EDGE, HSDPA, HSUPA, 3.5G, WLAN	S60 3rd Edition FP2	BB5.0	Candybar	5.0 megapixels	
Nokia E75	320 × 240 24-bit (16.7 million) Color	2009	P	GSM/EDGE/UMTS/3G/WLAN	S60 3rd Edition FP2	BB5.0	QWERTY Slider	3.2 megapixels	
Nokia E5-00	320 × 240 24-bit (16.7 million) Color	2010	P	GSM/EDGE/HSDPA/3G	S60 3rd Edition FP2	BB5.0	QWERTY Candybar	5.0 megapixels	
Nokia E7-00	360 × 640 24-bit (16.7 million) Color	2011	P	GSM/EDGE/HSDPA/3G	Symbian^3	*BB5.0*	QWERTY Slider candybar	8.0 megapixels	
Nokia E90 Communicator	800 × 352 24-bit (16.7 million) Color	2007	D	GSM/EDGE/3G/WLAN	S60 3rd Edition FP1	BB5.0	Clamshell	3.15 megapixels	

Nokia Nseries

The Nokia Nseries is Nokia's most advanced smartphone series. It is for people who wish to have advanced multimedia and connectivity features and as many other features as possible into one device.

Phone model	Screen type	Released	S.	Technology	Platform	Generation	Form factor	Camera
Nokia N70	176 × 208 18-bit (262,144) Color	2005	D	GSM/EDGE/UMTS	S60 2nd Edition	BB5.0	Candybar	2.0 megapixels
Nokia N70 Music Edition	176 × 208 18-bit (262,144) Color	2006	D	GSM/UMTS	S60 2nd Edition	BB5.0	Candybar	2.0 megapixels
Nokia N71	320 × 240 18-bit (262,144) Color	2006	D	GSM/UMTS	S60 3rd Edition	BB5.0	Clamshell	2.0 megapixels
Nokia N72	176 × 208 18-bit (262,144) Color	2006	D	GSM	S60 2nd Edition	BB5.0	Candybar	2.0 megapixels
Nokia N73	320 × 240 18-bit (262,144) Color	2006	D	GSM/EDGE/UMTS	S60 3rd Edition	BB5.0	Candybar	3.2 megapixels
Nokia N73 Music Edition	320 × 240 18-bit (262,144) Color	2006	D	GSM/EDGE/UMTS	S60 9.1 3rd Edition	BB5.0	Candybar	3.2 megapixels
Nokia N75	320 × 240 24-bit (16.7M) Color	2006	D	GSM/UMTS	S60 3rd Edition	BB5.0	Clamshell	2.0 megapixels
Nokia N76	320 × 240 24-bit (16.7M) Color	2007	D	GSM/UMTS	S60 3rd Edition	BB5.0	Clamshell	2.0 megapixels
Nokia N77	329 × 240 24-bit (16.7M) Color	2007	D	GSM/UMTS	S60 3rd Edition	BB5.0	Candybar	2.0 megapixels
Nokia N78	320 × 240 24-bit (16.7M) Color	2008	P	GSM EGPRS UMTS WLAN WCDMA HSDPA EGSM	S60 3rd Edition FP2	BB5.0	Candybar	3.2 megapixels
Nokia N79	320 × 240 24-bit (16.7M) Color	2008	P	GSM EGPRS UMTS WLAN WCDMA HSDPA EGSM	S60 3rd Edition	BB5.0	Candybar	5.0 megapixels
Nokia N8	360 × 640 (16.7M) Capacitive AMOLED touchscreen	2010	P	GSM EGPRS UMTS WLAN WCDMA HSDPA EGSM	Symbian^3	BB5.0	Candybar	12.0 megapixels (720p HD)
Nokia N80	352 × 416 18-bit (262,144) Color	2006	D	GSM/UMTS	S60 3rd Edition	BB5.0	Slide	3.15 megapixels
Nokia N81	320 × 240 24-bit (16.7M) Color	2007	D	GSM/UMTS/WLAN	S60 3rd Edition FP1	BB5.0	Slide	2.0 megapixels
Nokia N81 8GB	320 × 240 24-bit (16.7M) Color	2007	D	GSM/UMTS/WLAN	S60 9.2 3rd Edition FP1	BB5.0	Slide	2.0 megapixels
Nokia N82	320x240 24-bit (16.7M) Color	2007	D	GSM EGPRS UMTS WLAN WCDMA HSDPA EGSM	S60 3rd Edition FP1	BB5.0	Candybar	5.0 megapixels
Nokia N85	320x240 24-bit (16.7M) Color	2008	P	GSM EGPRS UMTS WLAN WCDMA HSDPA EGSM	S60 3rd Edition FP2	BB5.0	2-way Slide	5.0 megapixels
Nokia N86 8MP	320x240 24-bit (16.7M) Color	2009	P	GSM EGPRS UMTS WLAN WCDMA HSDPA EGSM	S60 3rd Edition FP2	BB5.0	Slide	8.0 megapixels
Nokia N9	854 × 480 (16.7M) Capacitive AMOLED touchscreen	2011	P	GSM EGPRS UMTS WLAN WCDMA HSDPA EGSM	MeeGo 1.2 Harmattan	BB5.0	Candybar	8.0 megapixels

Nokia N90	352 × 416 18-bit (262,144) Color	2005	D	GSM/UMTS	S60 2nd Edition	BB5.0	Clamshell	2.0 megapixels
Nokia N900	800 × 480 24-bit (16.7M) Resistive touchscreen	2009	P	GSM EGPRS UMTS WLAN WCDMA HSDPA EGSM	Maemo 5	BB5.0	Tablet	5.0 megapixels
Nokia N91	176 × 208 18-bit (262,144) Color	2005	D	GSM/UMTS	S60 3rd Edition	BB5.0	Slide	2.0 megapixels
Nokia N92	320 × 240 24-bit (16.7M) Color	2007	D	GSM/UMTS	S60 3rd Edition	BB5.0	Clamshell	2.0 megapixels
Nokia N93	320 × 240 18-bit (262,144) Color	2006	D	GSM/UMTS	S60 3rd Edition	BB5.0	Clamshell	3.2 megapixels
Nokia N93i	320 × 240 24-bit (16.7M) Color	2007	D	GSM EGPRS UMTS WLAN WCDMA HSDPA EGSM	S60 3rd Edition	BB5.0	Clamshell	3.2 megapixels
Nokia N95	320 × 240 24-bit (16.7M) Color	2006	D	GSM EGPRS UMTS WLAN WCDMA HSDPA EGSM	S60 3rd Edition FP1	BB5.0	2-way Slide	5.0 megapixels
Nokia N95 8GB	320 × 240 24-bit (16.7M) Color	2007	D	GSM EGPRS UMTS WLAN WCDMA HSDPA EGSM	S60 3rd Edition FP1	BB5.0	2-way Slide	5.0 megapixels
Nokia N950	854 × 480 24-bit (16.7M) Capacitive touchscreen	2011	U	GSM EGPRS UMTS WLAN WCDMA HSDPA EGSM	MeeGo 1.2 Harmattan	tbc	QWERTY keyboard, with tilt display	8.0 megapixels
Nokia N96	320 × 240 24-bit (16.7M) Color	2008	P	GSM EGPRS UMTS WLAN WCDMA HSDPA EGSM DVB-H	S60 3rd Edition FP2	BB5.0	2-way Slide	5.0 megapixels
Nokia N97	640 × 360 24-bit (16.7M) Color	2009	P	GSM EGPRS UMTS WLAN WCDMA HSDPA EGSM	S60 5th Edition	BB5.0	QWERTY keyboard, with tilt display	5.0 megapixels
Nokia N97 mini	360 × 640 24-bit (16.7M) Color	2009	P	GSM EGPRS UMTS WLAN WCDMA HSDPA EGSM	S60 5th Edition	BB5.0	QWERTY keyboard, with tilt display	5.0 megapixels

Note:

- Although part of the Nseries, the Nokia **N800** and **N810** Internet Tablets did not include phone functionality. See the Internet Tablets section.

Nokia Xseries

The Nokia Xseries targets a young audience with a focus on music and entertainment.

Phone model	Screen type	Released	S.	Technology	Platform	Generation	Form factor	Camera
Nokia X2-00	240 × 320 (256K) TFT Color	2010	P	GSM EDGE	Series 40 6th Edition	BB5.0	Candybar	5.0 megapixels
Nokia X2-01	320 × 240 (256K) TFT Color	2010	P	GSM EDGE	Series 40 6th Edition	BB5.0	QWERTY Candybar	VGA 640 × 480
Nokia X3-00	240 × 320 (256K) TFT Color	2009	P	GSM EDGE	Series 40 6th Edition	BB5.0	Slider	3.2 megapixels
Nokia X3-02 (Touch and Type)	240 × 320 (256K) TFT Color	2010	P	GSM EDGE, UMTS, WLAN	Series 40 6th Edition feature pack 1	BB5.0	Candybar	5.0 megapixels
Nokia X5	240 × 320 (16.7M) TFT Color (model-00) 320 × 240 (16.7M) TFT Color (model-01)	2010	P (model-00), (model-01)	TD-SCDMA (model-00) GSM, UMTS, WLAN (model-01)	Series 40 6th Edition (model-00) S60 3rd Edition FP2 (model-01)	BB5.0	Candybar (model-00) Slider (model-01)	5.0 megapixels
Nokia X6	640 × 360 (nHD) (16.7M) Color (Capacitive touchscreen)	2009 (Comes With Music/32GB) 2010 (16GB, 8 GB)	D (8 GB, 16GB, 32GB)	GSM EDGE UMTS WLAN	S60 5th Edition	BB5.0	Candybar	5.0 megapixels
Nokia X7-00	300 × 640 AMOLED (16.7M) Color (Capacitive touchscreen)	2011	P	GSM EDGE UMTS WLAN	Symbian^3 'Anna'(PR2.0)	BB5.0	Candybar	8 megapixels

Nokia Asha series

The Nokia Asha series is an affordable series optimized for social networking and sharing. Operating system used is Series 40.

Phone model	Screen type	Released	Technology	Platform	Generation	Form factor	Camera
Nokia Asha 200	320 x 240 pixels (256K)	2011 Q4	GSM GPRS EGPRS	Series 40 6th Edition feature pack 1	tbc	QWERTY Candybar	2.0 MP
Nokia Asha 201	320 x 240 pixels (256K)	2011 Q4	GSM GPRS EGPRS	Series 40 6th Edition feature pack 1	tbc	QWERTY Candybar	2.0 MP
Nokia Asha 300	240 x 320 pixels (256K)	2011 Q4	GSM WCDMA GPRS EGPRS HSDPA HSUPA	Series 40 6th Edition feature pack 1	tbc	Touch and Type Candybar	5.0 MP
Nokia Asha 303	240 x 320 pixels (256K)	2011 Q4	GSM WCDMA GPRS EGPRS HSDPA HSUPA WLAN	Series 40 6th Edition feature pack 1	tbc	Touch and Type QWERTY Candybar	3.2 MP

Nokia 200 is Dual-SIM phone

Nokia Lumia series

The Nokia Lumia series is a series running Windows Phone OS.

Phone model	Screen type	Released	S.	Technology	Platform	Generation	Form factor	Camera
Nokia Lumia 710	480x800 px 16m-color WVGA	2011	P	GSM EDGE	Windows Phone 7.5 (Mango)	BB5.0	Candybar	5.0 megapixels
Nokia Lumia 800	480x800 px 16m-color WVGA AMOLED (16.7M) Color (Capacitive touchscreen)	2011	P	GSM EDGE UMTS WLAN	Windows Phone 7.5 (Mango)	BB5.0	Candybar	8.0 megapixels
Nokia Lumia 900	480x800 px 16m-color WVGA AMOLED (16.7M) Color (Capacitive touchscreen)	2011	P	GSM EDGE UMTS WLAN	Windows Phone 7.5 (Mango)	BB5.0	Candybar	8.0 megapixels

Other phones

Nokia N-Gage – Mobile gaming devices (discontinued)

Phone model	Screen type	Released	S.	Technology	Generation	Form factor
N-Gage	176 × 208 12-bit (4096) Color	2003	D	GSM	DCT4	Candybar
N-Gage QD	176 × 208 12-bit (4096) Color	2004	D	GSM	DCT4	Candybar

Cardphones (PCMCIA)

Phone model	Released	S.	Technology	Generation	Form factor
Cardphone	2001	D	GSM	DCT3	PC card
Cardphone 2.0	2001	D	GSM	DCT3	PC card
D211	2003	D	GSM/WLAN	DCT4	PC card
D311	2003	P	GSM/WLAN	DCT4	PC card

Concept phones

Nokia developed a phone concept, never realised as a working device, in the 2008 Nokia Morph.

Services and solutions

Online services

- Ovi (Nokia)
- N-Gage (service)

Software solutions

- Nokia PC Suite
- Nokia Ovi Suite
- Nokia Photos
- Nokia Lifeblog
- Intellisync Mobile Suite
- Nokia Business Center
- Nokia Sensor
- Gammu and Wammu for both Linux and MS Windows[3]
- Nokia Maps
- Nokia Sports Tracker
- Nokia Software Updater
- Qt (framework)

Security solutions

IP appliances run IPSO FreeBSD based operating system, work with Check Point's firewall and VPN products. Nokia Network Voyager is an SSL-secured, Web-based element management interface.

- Nokia IP 40
- Nokia IP 130
- Nokia IP 260
- Nokia IP 265
- Nokia IP 330
- Nokia IP 350
- Nokia IP 380
- Nokia IP 390 (EU Only)
- Nokia IP 530
- Nokia IP 710
- Nokia IP 1220
- Nokia IP 1260
- Nokia IP 2250
- Nokia Horizon Manager

In 2004, Nokia began offering their own SSL VPN appliances based on IP Security Platforms and the pre-hardened Nokia IPSO operating system. Client integrity scanning and endpoint security technology was licensed from Positive Networks.[4]

- Nokia 50s
- Nokia 105s
- Nokia 500s

Other products

Mini laptops

On August 24, 2009, Nokia announced that they will be re-entering the PC business with a high-end mini laptop called the Nokia Booklet 3G.[5]

Internet Tablets

Nokia's Internet Tablets were designed for wireless Internet browsing and e-mail functions and did not include phone capabilities. The Nokia N800 and N810 Internet Tablets were also marketed as part of Nseries. See the Nseries section.

- Nokia 770 Internet Tablet
- Nokia N800 Internet Tablet
- Nokia N810 Internet Tablet
- Nokia N810 WiMAX Edition

The Nokia N900, the successor to the N810, has phone capabilities and is not officially marketed as an Internet Tablet, but rather as an actual Nseries smartphone.

GPS products

- Nokia 5140 GPS Cover
- Nokia wireless GPS module LD-1W
- Nokia wireless GPS module LD-3W
- Nokia Bluetooth GPS Module LD-4W
- Navigation Kit for Nokia 770 Internet Tablet, including LD-3W GPS receiver and software
- Nokia 330 Navigator, that supports an external TMC module.
- Nokia 500 Navigator

Accessories

Nokia produces accessories to their products too many to list here. Such accessories include:

- **Carrying and styling:** carrying cases, phone jewellery, shells
- **Car solutions:** car kits, car phones, portable solutions, mobile holders, car accessories, car navigation
- **Headsets:** audio adapters, bluetooth headsets, wired headsets, loopsets
- **Memory cards and cables**
- **Music related products:** audio adapters, music packs, music streaming, speakers
- **Navigation:** GPS modules, navigation kits, car navigation
- **Home and office:** desk stands, imaging, wireless digital pens, wireless keyboards, mobile TV receivers
- **Power:** batteries, chargers, charger adapters
- **Internet Stick:** Nokia HSPA-modem CS-15 operates in 3G network, with maximum download speeds of 14.4 Mbit/s and upload speeds of 2.1 Mbit/s.

Products marketed by Nokia Siemens Networks

Nokia Siemens Networks provides wireless and wired network infrastructure, communications and networks service platforms, as well as professional services to operators and service providers. The company is jointly owned by Nokia and Siemens AG and consolidated by Nokia.[6]

Telephone switches

- Nokia DX 200
- Nokia DX 220
- Nokia DX 220 Compact

TETRA

Nokia sold its PMR business to Cassidian, a division of EADS.

- Nokia THR400
- Nokia TMR400 (car model)
- Nokia THR420
- Nokia THR600
- Nokia THR850
- Nokia THR880
- Nokia TMR880 (car model)
- Nokia THR880i (external link) [7]

Products marketed by Vertu, Nokia's luxury phones brand

Vertu is an independent company, owned by Nokia, that manufactures exclusively hand crafted high-end mobile phones.

Phone model	Screen type	Released	S.	Technology	Generation	Form factor
Signature	116 × 148 12-bit (4096) Color	2002		GSM	Unknown	Candybar
Diamond	?	?		?	Unknown	Candybar
Ascent	116 × 148 12-bit (4096) Color	2004		GSM	Unknown	Candybar
Constellation-	116 × 148 12-bit (4096) Color	2004		GSM	Unknown	Candybar
Signature S Design	Unknown	2008		GSM|UMTS|WLAN	Unknown	Candybar

Materials used include platinum, 18 carat white gold, 18 carat yellow gold, stainless steel, ruby keypad bearings, and a sapphire crystal display for the "Signature" and leather, 316L Surgical stainless steel and Liquidmetal for the "Ascent".

Discontinued product lines

Personal computers

In the 1980s, Nokia's personal computer division Nokia Data manufactured a series of personal computers by the name of MikroMikko. Nokia's PC division was sold to the British computer company ICL in 1991, which later became part of Fujitsu.[8] Nokia re-entered the PC market in 2009 with the introduction of the Nokia Booklet 3G mini laptop.

Computer displays

Nokia was also known for producing very high quality CRT and early TFT LCD Multigraph displays for PC and larger systems application. The Nokia Display Products' branded business was sold to ViewSonic in 2000.[9]

Digital television

Nokia used to manufacture digital set-top boxes.

- Nokia DBox
- Nokia DBox2
- Nokia Mediamaster 9500 S
- Nokia Mediamaster 9500 C
- Nokia Mediamaster 9600 S
- Nokia Mediamaster 9600 C
- Nokia Mediamaster 9610 S
- Nokia Mediamaster 9800 S
- Nokia Mediamaster 9850 T
- Nokia Mediamaster 9900 S
- Nokia Mediamaster 110 T
- Nokia Mediamaster 210 T
- Nokia Mediamaster 221 T
- Nokia Mediamaster 230 T
- Nokia Mediamaster 260 T
- Nokia Mediamaster 260 C
- Nokia Mediamaster 310 T

ADSL modems

- Nokia M10
- Nokia M11
- Nokia M1122
- Nokia MW1122
- Nokia M5112
- Nokia M5122
- Nokia Ni200[10]
- Nokia Ni500[11]

WLAN products

- Nokia A020 WLAN access point
- Nokia A021 WLAN access point/router
- Nokia A032 WLAN access point[12]
- Nokia C020 PC card IEEE 802.11 2 Mbit/s, DSSS (produced by Samsung)
- Nokia C021 PC card, with external antenna
- Nokia C110 PC card IEEE 802.11b 11 Mbit/s
- Nokia C111 PC card, with external antennas
- Nokia MW1122 ADSL modem with wireless interface
- Nokia D211 WLAN/GPRS PC card

Military communications and equipment

Nokia has developed the Sanomalaitejärjestelmä ("Message device system") for the Finnish Defence Forces. It includes:

- Sanomalaite M/90
- Partiosanomalaite
- Keskussanomalaite

Nokia also manufactured the M61 gas mask for the Finnish Defence Forces.

Past industries

In the past, Nokia has produced at one time or another:[1]

- Paper products
- Tires (car and bicycle)
- Footwear (including Wellington boots)
- Communications cables
- Consumer electronics such as televisions
- Personal computers
- Electricity generation machinery
- Military technology and equipment
- Robotics
- Capacitors
- Plastics
- Aluminium
- Chemicals

See also

- Nokian Tyres
- List of Sony Ericsson products
- List of Motorola products

References

[1] "Nokia – Towards Telecommunications" (http://www.nokia.com/NOKIA_COM_1/About_Nokia/Sidebars_new_concept/Broschures/TowardsTelecomms.pdf) (PDF). Nokia Corporation. August 2000. . Retrieved 2009-08-27.

[2] here (http://blogs.s60.com/seeintos60/2008/10/s60_5th_edition_and_the_nokia.html)

[3] http://wammu.eu/

[4] Business Wire (2004). "Nokia Licenses Remote Access Technology from Positive Networks" (http://www.tmcnet.com/usubmit/2004/oct/1083797.htm). TMCnet.com. . Retrieved 2006-11-13.

[5] "Nokia Booklet 3G brings all day mobility to the PC world" (http://www.nokia.com/press/press-releases/showpressrelease?newsid=1336683) (Press release). Nokia Corporation. 2009-08-24. . Retrieved 2009-08-26.

[6] "Nokia Siemens Networks starts operations and assumes a leading position in the communications industry" (http://www.nokia.com/A4136002?newsid=1116423) (Press release). Nokia Corporation. 2007-04-02. . Retrieved 2009-04-07.

[7] http://www.nwr.nokia.com/BaseProject/Sites/NOKIA_MAIN_18022/CDA/Categories/Operators/Government/Library/_Content/_Static_Files/tetra_touch_03_2004_gov.pdf

[8] "Historia: 1991–1999" (http://www.fujitsu.com/fi/about/history/1991/) (in Finnish). Fujitsu Services Oy, Finland. . Retrieved 2010-09-22.

[9] "ViewSonic Corporation Acquires Nokia Display Products' Branded Business" (http://press.nokia.com/PR/200001/775025_5.html) (Press release). Nokia Corporation. 2000-01-17. . Retrieved 2009-03-22.

[10] http://dsl.nokia.co.nz/ni200/index.htm

[11] http://dsl.nokia.co.nz/ni500/index.htm

[12] http://www.europe.nokia.com/nokia/0,,2375,00.html

External links

- Nokia - Phone Software Update (http://www.nokia.co.uk/nokia/0,1522,,00.html?orig=/softwareupdate)
- All Nokia DCT and BB generation list (http://www.nokia-tuning.net/index.php?s=dct)
- Nokia museum (over 300 phones) (http://www.nokiamuseum.com)
- Nokia graphical timeline (1982 to 2006) (http://www.newlaunches.com/entry_images/1107/12/nokia_timeline.php)
- Nokia Mobile Reviews (http://www.mobilespect.com/category/nokia)

Smartphone

A **smartphone** is a high-end mobile phone built on a mobile computing platform, with more advanced computing ability and connectivity than a contemporary feature phone.[1] [2] [3] The first smartphones were devices that mainly combined the functions of a personal digital assistant (PDA) and a mobile phone or camera phone. Today's models also serve to combine the functions of portable media players, low-end compact digital cameras, pocket video cameras, and GPS navigation units. Modern smartphones typically also include high-resolution touchscreens, web browsers that can access and properly display standard web pages rather than just mobile-optimized sites, and high-speed data access via Wi-Fi and mobile broadband.

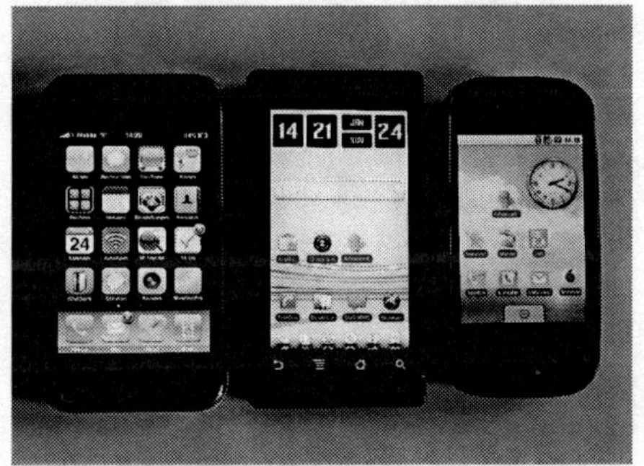

Modern smartphones.

The most common mobile operating systems (OS) used by modern smartphones include Google's Android, Apple's iOS, Microsoft's Windows Phone, Nokia's Symbian, RIM's BlackBerry OS, and embedded Linux distributions such as Maemo and MeeGo. Such operating systems can be installed on many different phone models, and typically each device can receive multiple OS software updates over its lifetime.

The distinction between smartphones and feature phones can be vague and there is no official definition for what constitutes the difference between them. One of the most significant differences is that the advanced application programming interfaces (APIs) on smartphones for running third-party applications[4] can allow those applications to have better integration with the phone's OS and hardware than is typical with feature phones. In comparison, feature phones more commonly run on proprietary firmware, with third-party software support through platforms such as Java ME or BREW.[1] An additional complication in distinguishing between smartphones and feature phones is that over time the capabilities of new models of feature phones can increase to exceed those of phones that had been promoted as smartphones in the past.

History

Early years

The first smartphone was the IBM Simon; it was designed in 1992 and shown as a concept product[5] that year at COMDEX, the computer industry trade show held in Las Vegas, Nevada. It was released to the public in 1993 and sold by BellSouth. Besides being a mobile phone, it also contained a calendar, address book, world clock, calculator, note pad, e-mail client, the ability to send and receive faxes, and games. It had no physical buttons, instead customers used a touchscreen to select telephone numbers with a finger or create faxes and memos with an optional stylus. Text was entered with a unique on-screen "predictive" keyboard. By today's standards, the Simon would be a fairly low-end product, lacking a camera and the ability to download third-party applications. However, its feature set at the time was highly advanced.

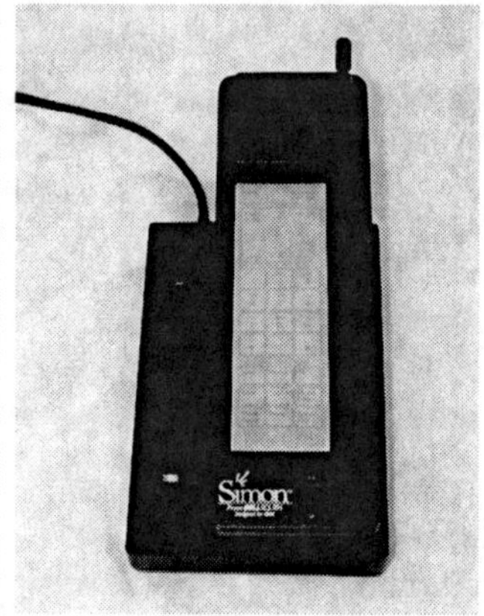

IBM Simon (introduced 1992) shown in the charging station

The Nokia Communicator line was the first of Nokia's smartphones starting with the Nokia 9000, released in 1996. This distinctive palmtop computer style smartphone was the result of a collaborative effort of an early successful and costly personal digital assistant (PDA) by Hewlett-Packard combined with Nokia's best-selling phone around that time, and early prototype models had the two devices fixed via a hinge. The Communicators are characterized by a clamshell design, with a feature phone display, keyboard and user interface on top of the phone, and a physical QWERTY keyboard, high-resolution display of at least 640×200 pixels and PDA user interface under the flip-top. The software was based on the GEOS V3.0 operating system, featuring email communication and text-based web browsing. In 1998, it was followed by Nokia 9110, and in 2000 by Nokia 9110i, with improved web browsing capability.

In 1997 the term 'smartphone' was used for the first time when Ericsson unveiled the concept phone GS88,[6] [7] the first device labelled as 'smartphone'.[8]

Symbian

The Nokia 9210 Communicator (Symbian 2000 model smartphone)

In 2000, the touchscreen Ericsson R380 Smartphone was released.[9] It was the first device to use an open operating system, the Symbian OS.[10] It was the first device marketed as a 'smartphone'.[11] It combined the functions of a mobile phone and a personal digital assistant (PDA).[12] In December 1999 the magazine Popular Science appointed the Ericsson R380 Smartphone to one of the most important advances in science and technology.[13] It was a groundbreaking device since it was as small and light as a normal mobile phone.[14] In 2002 it was followed up by P800.[15]

Also in 2000, the Nokia 9210 communicator was introduced, which was the first color screen model from the Nokia Communicator line. It was a true smartphone with an open operating system, the Symbian OS. It was followed by the 9500 Communicator, which also was Nokia's first cameraphone and first Wi-Fi phone. The 9300 Communicator was smaller, and the latest E90 Communicator includes GPS. The Nokia Communicator model is remarkable for also having been the most costly phone model sold by a major brand for almost the full life of the model series, costing easily 20% and sometimes 40% more than the next most expensive smartphone by any major producer.

In 2007 Nokia launched the Nokia N95 which integrated a wide range of multimedia features into a consumer-oriented smartphone: GPS, a 5 megapixel camera with autofocus and LED flash, 3G and Wi-Fi connectivity and TV-out. In the next few years these features would become standard on high-end smartphones. The Nokia 6110 Navigator is a Symbian based dedicated GPS phone introduced in June 2007.

In 2010 Nokia released the Nokia N8 smartphone with a stylus-free capacitive touchscreen, the first device to use the new Symbian^3 OS.[16] It featured a 12 megapixel camera with Xenon flash able to record HD video in 720p, described by Mobile Burn as the best camera in a phone,[17] and satellite navigation that Mobile Choice described as the best on any phone.[18] It also featured a front-facing VGA camera for videoconferencing.

Symbian was the number one smartphone platform by market share from 1996 until 2011 when it dropped to second place behind Google's Android OS. In February 2011, Nokia announced that it would replace Symbian with Windows Phone as the operating system on all of its future smartphones.[19] This transition was completed in October 2011, when Nokia announced its first line of Windows Phone 7.5 smartphones, Lumia 710 and 800.[20]

Palm, Windows, and BlackBerry

In the late 1990s the vast majority of mobile phones had only basic phone features and many people who needed functionality beyond that also carried PDA and/or pager type devices running early versions of operating systems such as Palm OS, BlackBerry OS or Windows CE/Pocket PC.[1] Later versions of these systems started integrating cell phone capabilities with their PDA and messaging features and support of third-party applications. Today, high-end devices running these systems are often branded smartphones.

The HTC Touch Pro2 smartphone (May 2009)

In early 2001, Palm, Inc. introduced the Kyocera 6035, the first smartphone to be deployed in widespread use in the United States. This device combined the features of a personal digital assistant (PDA) with a wireless phone that operated on the Verizon Wireless network. For example, a user could select a name from the PDA contact list, and the device would dial that contact's phone number. The device also supported limited web browsing.[21] The device received a very positive reception from technology publications, but the product line never became widespread outside North America.[22]

In 2001 Microsoft announced its Windows CE Pocket PC OS would be offered as "Microsoft Windows Powered Smartphone 2002."[23] Microsoft originally defined its Windows Smartphone products as lacking a touchscreen and offering a lower screen resolution compared to its sibling Pocket PC devices.

In early 2002 Handspring released the Palm OS Treo smartphone, utilizing a full keyboard that combined wireless web browsing, email, calendar, and contact organizer with mobile third-party applications that could be downloaded or synced with a computer.[24]

In 2002 RIM released their first BlackBerry devices with integrated phone functionality and shifted the positioning of their products from 2-way pagers to email-capable mobile phones. The BlackBerry line evolved into the first smartphone optimized for wireless email use and had achieved a total customer base of about 32 million subscribers by December 2009.[25]

In February 2011 Nokia announced a plan to make Microsoft Windows Phone its main operating system of choice for new Nokia smartphones.[19]

iPhone

The original iPhone (June 2007)

In 2007, Apple Inc. introduced its first iPhone. It was initially costly, priced at $499 for the cheaper of two models on top of a two year contract. The first mobile phone to use a multi-touch interface, the iPhone was notable for its use of a large touchscreen for direct finger input as its main means of interaction, instead of having a stylus, keyboard, and/or keypad, which were the typical input methods for other smartphones at the time. The iPhone featured a web browser that *Ars Technica* then described as "far superior" to anything offered by that of its competitors.[26] Initially lacking the capability to install native applications beyond the ones built-in to its OS, at WWDC in June 2007 Apple announced that the iPhone would support third-party "web 2.0 applications" running in its web browser that share the look and feel of the iPhone interface.[27] As a result of the iPhone's initial inability to install third-party native applications, some reviewers did not consider the originally released device to accurately fit the definition of a smartphone "by conventional terms."[28] A process called jailbreaking emerged quickly to provide unofficial third-party native applications. The different functions of the iPhone (including a GPS unit, kitchen timer, radio, map book, calendar, notepad, and many others) allowed consumers to replace all of these items.[29]

In July 2008, Apple introduced its second generation iPhone with a lower list price starting at $199 and 3G support. Released with it, Apple also created the App Store, adding the capability for any iPhone or iPod Touch to officially execute additional native applications (both free and paid) installed directly over a Wi-Fi or cellular network, without the more typical process at the time of requiring a PC for installation. Applications could additionally be browsed through and downloaded directly via the iTunes software client on Macintosh and Windows PCs, rather than by searching through multiple sites across the Internet. Featuring over 500 applications at launch,[30] Apple's App Store was immediately very popular,[31] quickly growing to become a huge success.[32] [33]

In June 2010, Apple introduced iOS 4, which included APIs to allow third-party applications to multitask,[34] and the iPhone 4, which included a 960×640 pixel display with a pixel density of 326 pixels per inch (ppi), a 5 megapixel camera with LED flash capable of recording HD video in 720p at 30 frames per second, a front-facing VGA camera for videoconferencing, a 1 GHz processor, and other improvements.[35] In early 2011 the iPhone 4 became available through Verizon Wireless, ending AT&T's exclusivity of the handset in the U.S.,[36] [37] [38] and allowing the handset's 3G connection to be used as a wireless Wi-Fi hotspot for the first time, to up to 5 other devices.[39] Software updates subsequently added this capability to other iPhones running iOS 4.[40] [41]

The iPhone 4S was announced on October 4, 2011, improving upon the iPhone 4 with a dual core A5 processor, an 8 megapixel camera capable of recording 1080p video at 30 frames per second, World phone capability allowing it to work on both GSM & CDMA networks, and the Siri automated voice assistant.[42] On October 10, Apple announced that over one million iPhone 4Ss had been pre-ordered within the first 24 hours of it being on sale, beating the 600,000 device record set by the iPhone 4,[43] [44] despite the iPhone 4S failing to impress some critics at the announcement[45] [46] due to their expectations of an "iPhone 5" with rumored drastic changes compared to the iPhone 4 such as a new case design and larger screen.[47] Along with the iPhone 4S Apple also released iOS 5 and iCloud, untethering iOS devices from Macintosh or Windows PCs for device activation, backup, and synchronization,[48] along with additional new and improved features.[49]

There are about 35 percent of Americans that have some sort of smartphone. This shows that the market is spreading fast and there are also more capabilities for smartphones because of this spread.[50]

Smartphones are also mainly valuable based on the operating system. For example, the iPhone runs on the iOS and other devices run different operating systems which makes the functionality of these systems different.[51]

Android

The Android operating system for smartphones was released in 2008. Android is an open-source platform backed by Google, along with major hardware and software developers (such as Intel, HTC, ARM, Motorola and Samsung, to name a few), that form the Open Handset Alliance.[52] The first phone to use Android was the HTC Dream, branded for distribution by T-Mobile as the G1. The software suite included on the phone consists of integration with Google's proprietary applications, such as Maps, Calendar, and Gmail, and a full HTML web browser. Android supports the execution of native applications and a preemptive multitasking capability (in the form of services). Third-party apps are available via the Android Market (released October 2008), including both free and paid apps.

Galaxy Nexus, the latest "Google phone"

In January 2010, Google launched the Nexus One smartphone using its Android OS. Although Android has multi-touch abilities, Google initially removed that feature from the Nexus One,[53] but it was added through a firmware update on February 2, 2010.[54]

Concerning the Xperia Play smartphone, an analyst at CCS Insight said in March 2011 that "Console wars are moving to the mobile platform".[55] In the same month, the HTC EVO 3D was announced by HTC Corporation, which can produce 3D effects with no need for special glasses (autostereoscopy).[56] The HTC EVO 3D was officially released on June 24, 2011.[57]

Bada

The Bada operating system for smartphones was announced by Samsung on 10 November 2009.[58] [59] The first Bada-based phone was the Samsung Wave S8500, released on June 1, 2010,[60] [61] which sold one million handsets in its first 4 weeks on the market.[62]

Samsung shipped 3.5 million phones running Bada in Q1 of 2011.[63] This rose to 4.5 million phones in Q2 of 2011.[64]

Patent licensing and litigation

Recently the number of lawsuits, trade complaints, and countersuits and complaints based on patents and designs in the markets for smartphones, and devices based on smartphone OSes such as Android, has been increasing significantly.

Timeline[65] [66] [67] [68] [69] (initial suits, countersuits, rulings, licence agreements, and other major events in *italics*):

- 2009, Oct 22: *Nokia sues Apple over 10 patents.*[70] [71]
- 2009, Dec 11: *Apple countersues Nokia over 13 patents.*[72]
- 2009, Dec 29: Nokia files a second lawsuit[73] and a U.S. International Trade Commission (ITC) complaint against Apple over 7 more patents.[74]

- 2010, Jan 15: Apple files an ITC complaint against Nokia over 9 patents.[75] [76]
- 2010, Feb 19: Apple drops 4 patents from their countersuit against Nokia that are in their ITC complaint against Nokia.
- 2010, Feb 24: Apple countersues Nokia in Nokia's second lawsuit, over the 9 patents that are in Apple's ITC complaint.
- 2010, Mar 02: *Apple sues HTC over 10 patents and files an ITC complaint against HTC over 10 other patents.*[77] [78] [79] [80] [81]
- 2010, Apr 26: 5 of the patents in Apple's ITC complaint against Nokia are merged into their ITC complaint against HTC.
- 2010, Apr 27: *HTC signs an agreement with Microsoft to licence Microsoft patents in return for royalties on HTC's Android-based devices*[82] [83] (rumored to be $5 per handset).
- 2010, May 7: Nokia files a third lawsuit against Apple over 5 more patents.[84]
- 2010, May 12: *HTC files an ITC complaint against Apple over 5 patents.*[85]
- 2010, May 28: S3 Graphics files an ITC complaint against Apple over 4 patents used in the iPhone, iPod Touch, iPad, and Apple computers.[86]
- 2010, Jun 28: Apple countersues Nokia in Nokia's third lawsuit, over 7 more patents.
- 2010, Jul 06: *HTC countersues Apple over 3 patents.*
- 2010, Jul 21: Nokia drops 1 patent from their ITC complaint against Apple.
- 2010, Aug 12: *Oracle sues Google over 7 patents relating to the use of Java in Android.*[87]
- 2010, Sep 17: Nokia adds 2 more patents to their third lawsuit against Apple.
- 2010, Sep 27: Apple sues Nokia in the UK and Germany over 9 patents.
- 2010, Sep 30: Nokia countersues Apple in Germany over 4 patents.
- 2010, Oct 01: *Microsoft files an ITC complaint and a lawsuit against Motorola over 9 patents.*[88] [89]
- 2010, Oct 06: *Motorola sues Apple over 18 patents, and files an ITC complaint against Apple over 6 of them.*[90]
- 2010, Oct 08: *Motorola files a request for declaratory judgement that they do not infringe 12 Apple patents, and that those patents be declared invalid.*[91] [92]
- 2010, Oct 12: Nokia adds 3 more patents to their countersuit against Apple in Germany.
- 2010, Oct 25: Nokia sues Apple in another German court over 5 patents.
- 2010, Oct 28: Apple drops 4 patents from their ITC complaint against HTC and/or Nokia.
- 2010, Oct 29: *Apple sues Motorola over 6 patents, and files an ITC complaint against Motorola over 3 of them.*[93] [94]
- 2010, Nov 05: HTC drops 1 patent from their ITC complaint against Apple.
- 2010, Nov 09: *Microsoft alleges Motorola has failed to comply with RAND (reasonable and non-discriminatory) licensing obligations.*
- 2010, Nov 10: *Motorola sues Microsoft over 7 patents in one court and 9 patents in another.*
- 2010, Nov 18: *Apple makes counterclaims against Motorola over 6 patents.*
- 2010, Nov 22: *Motorola files an ITC complaint against Microsoft over 5 patents.*
- 2010, Dec 01: *Apple adds the 12 patents to their suit against Motorola that Motorola previously requested declaratory judgement that they do not infringe.*[95]
- 2010, Dec 03: Nokia countersues Apple in the UK over 4 patents, and files a new suit against Apple in the Netherlands over 2 patents.
- 2010, Dec 03: Apple countersues Nokia in Nokia's second German lawsuit, over 1 patent and 2 utility models.
- 2010, Dec 06: Nokia drops 1 patent from their ITC complaint against Apple.
- 2010, Dec 15 and 22: Nokia and Apple take their first German suit/countersuit to the Federal Patent Court of Germany.
- 2010, Dec 23: Motorola files a third lawsuit against Microsoft over 3 patents.
- 2010, Dec 23: Microsoft countersues Motorola over 7 patents.

- 2011, Jan 06: The third Nokia/Apple lawsuit/countersuit is transferred to the location of the first and second ones.
- 2011, Jan 18: Apple seeks to invalidate one Nokia patent in the UK, which it was not yet being sued over.
- 2011, Jan 18: Motorola drops 1 patent from their lawsuits against Microsoft.
- 2011, Jan 19: Microsoft counterclaims against Motorola, asserting 5 patents.
- 2011, Jan 25: Microsoft counterclaims against Motorola, asserting 2 patents.
- 2011, Feb 14: Motorola adds 2 patents to their lawsuits against Microsoft.
- 2011, Feb 22: Apple drops 1 more patent from their ITC complaint against HTC and Nokia.
- 2011, Mar 21: *Microsoft sues Barnes & Noble over the Android operating system in the Nook ebook reader.*[96]
- 2011, Mar 25: *ITC finds that Apple does not infringe on 5 Nokia patents.*
- 2011, Mar 29: Nokia files an ITC complaint against Apple over 7 more patents, and a fourth lawsuit over 6 of those.[97] [98]
- 2011, Apr 15: *Apple sues Samsung for patent and trademark infringement* (7 utility patents, 3 design patents, 3 registered trade dresses, 6 trademarked icons) with its Galaxy line of mobile products, including the *Galaxy S* smartphone and the *Galaxy Tab* tablet.[99] [100]
- 2011, Apr 22: *Samsung sues Apple in South Korea (5 patents), Japan (2 patents), and Germany (3 patents).*[101]
- 2011, Apr 28: *Samsung countersues Apple over 10 patents.*[102]
- 2011, Apr 29: Apple drops 1 more patent from their ITC complaint against HTC.
- 2011, May 18: *Samsung ordered to provide Apple samples of the announced Galaxy S2, Infuse 4G,* and *Infuse 4G LTE* smartphones, as well as the *Galaxy Tab 8.9* and *Galaxy Tab 10.1* tablets as part of Apple's lawsuit against the company.[103] [104]
- 2011, May 18: *Samsung files a court motion for Apple to provide samples of the unannounced iPhone 5 and iPad 3 prototypes.*[105]
- 2011, Jun 14: *Nokia and Apple settle their litigation with Apple agreeing to pay Nokia an undisclosed one-time payment as well as continuing royalties.*[106] [107]
- 2011, Jun 16: *Apple amends its lawsuit against Samsung,* dropping 2 utility patents and 1 design patent, and adding 3 new utility patents plus 4 trade dress applications, *now covering the Samsung Galaxy Tab 10.1*[108]
- 2011, Jun 22: Apple countersues Samsung in South Korea over an unknown number of patents.
- 2011, Jun 22: *Samsung's motion to be provided samples of Apple's unannounced iPhone 5 and iPad 3 prototypes is denied.*[109]
- 2011, Jun 27: General Dynamics Itronix signs an agreement with Microsoft to licence Microsoft patents in return for royalties on General Dynamics Itronix's Android-based devices.[110] [111]
- 2011, Jun 28: Samsung files an ITC complaint and a lawsuit against Apple over 5 patents.
- 2011, Jun 29: Samsung sues Apple in London, UK over an unknown number of patents, and a Samsung lawsuit against Apple in Italy becomes known (details unknown).
- 2011, Jun 29: Velocity Micro signs an agreement with Microsoft to licence Microsoft patents in return for royalties on Velocity Micro's Android-based devices.[111] [112]
- 2011, Jun 30: Samsung converts its countersuit against Apple into counterclaims against Apple's suit, dropping 2 patents but adding 4 more.
- 2011, Jun 30: *A consortium of companies made up of Apple, EMC Corporation, Ericsson, Microsoft, Research In Motion and Sony win against Google*[113] *in an auction of over 6,000 Nortel mobile-related telecommunications patents for $4.5 billion USD.*[114] [115]
- 2011, Jun 30: Onkyo signs an agreement with Microsoft to licence Microsoft patents in return for royalties on Onkyo's Android-based devices.[111] [116]
- 2011, Jul 01: *Apple files for preliminary injunction against 4 Samsung products: Infuse 4G, Galaxy S 4G, Droid Charge,* and *Galaxy Tab 10.1* based on 3 design patents and 1 utility patent.[117]
- 2011, Jul 01: *ITC rules that Apple infringes on 2 patents held by S3 Graphics, while not infringing on 2 others.*[118]

- 2011, Jul 05: *Apple files an ITC complaint against Samsung over 6 smartphones and 2 tablets* infringing 5 utility patents and 2 design patents.
- 2011, Jul 05: Wistron signs an agreement with Microsoft to licence Microsoft patents in return for royalties on Wistron's Android-based devices.[111] [119] [120]
- 2011, Jul 06: *HTC agrees to purchase S3 Graphics* to secure 235 patents for use in its defense against Apple.[121] [122] [123]
- 2011, Jul 06: *Microsoft seeks $15 licensing fees from Samsung for a range of claimed patent violations on every Android device.*[124]
- 2011, Jul 11: Apple files a second ITC complaint against HTC over 5 more patents, and sues HTC over 4 patents from this second ITC complaint that they weren't already suing HTC over.[125] [126]
- 2011, Jul 11-12: Google acquires 1,029 Patents from IBM for an undisclosed amount.[127] [128]
- 2011, Jul 15: *ITC finds HTC infringes on 2 Apple patents.*[129]
- 2011, Jul 29: HTC sues Apple in London, UK over an unknown number of patents.
- 2011, Aug 02: *Apple sues Samsung in Australia over 10 patents, resulting in Samsung delaying the launch and halting advertising of the Samsung Galaxy Tab 10.1 tablet in Australia* to an indefinite date.[130] [131]
- 2011, Aug 09: *A German court issues a preliminary injunction against the Samsung Galaxy Tab 10.1* in Apple's lawsuit against Samsung *which causes its sale to be banned in most of Europe.*[132] [133]
- 2011, Aug 15: *Google announces its intention to buy Motorola Mobility for $12.5 billion USD.* Eighteen of Motorola's patents could potentially be used for defense or countersuits against Apple and Microsoft, and may influence the smartphone war. These patents may change the balance of power, and force the various players to settle their lawsuits.[134] [135]
- 2011, Aug 16: *The Samsung Galaxy Tab 10.1 sales ban in Europe is lifted outside of Germany.*[136] [137]
- 2011, Aug 17: Google acquires 1,023 more patents from IBM for an undisclosed amount (not revealed until 13 Sep 2011).[138]
- 2011, Aug 23: *Microsoft files a complaint with the ITC requesting a ban on several key Motorola smartphones and devices in the USA* based on infringements of 7 patents.[139] [140]
- 2011, Aug 24: *A court in the Netherlands rules that Samsung will be banned from selling the Galaxy S, Galaxy S II and Galaxy Ace in a number of European countries* due to Apple's patent infringement claims.[141]
- 2011, Sep 02: *Apple granted preliminary injunction against Samsung preventing display of the prototype Samsung Galaxy Tab 7.7 tablet at the IFA trade show in Berlin.*[142]
- 2011, Sep 02: *Apple court filings assert that Andy Rubin got inspiration for Android framework while working at Apple* before working at General Magic and Danger, Inc.[143]
- 2011, Sep 07: *HTC countersues Apple using nine patents from Google.* The move is seen as a possible first step for Google giving direct support in lawsuits involving manufacturers using Android.[144] [145] [146] [147]
- 2011, Sep 08: *Acer*[148] *and ViewSonic*[149] *sign patent license agreements with Microsoft* regarding their use of Android on smartphones and tablets.[150] [151]
- 2011, Sep 09: *Apple's preliminary injunction against sales of the Samsung Galaxy Tab 10.1 in Germany is upheld.*[152]
- 2011, Sep 12: Samsung announces a lawsuit against Apple in France that had been filed in July over 3 patents.[153]
- 2011, Sep 12: Apple countersues Samsung in the UK over an unknown number of patents.[154]
- 2011, Sep 13: Google's August 17 acquisition of 1,023 patents from IBM is revealed by the U.S. Patent and Trademark Office.[138] [155]
- 2011, Sep 17: Samsung countersues Apple in Australia over 7 patents.[156]
- 2011, Sep 28: *Samsung signs an agreement with Microsoft to licence Microsoft patents in return for royalties on Samsung's Android-based devices.*[157] [158] [159]

- 2011, Oct 12: *An Australian court issues a preliminary injunction against the Samsung Galaxy Tab 10.1 in Apple's lawsuit against Samsung which prevents its sale in Australia leading up to the 2011 holiday season.*[160]
- 2011, Oct 13: *Quanta signs an agreement with Microsoft to licence Microsoft patents in return for royalties on Quanta's Android and Chrome-based devices.*[161] [162]
- 2011, Oct 13: *Judge in Apple's U.S. lawsuit against Samsung agrees that Samsung's tablets infringe on Apple's patents, but also that the validity of some of the patents might be questionable.*[163]

Screen

Screens on smartphones vary largely in both display size and display resolution. The most common screen sizes range from 2 inches to over 4 inches (measured diagonally). Some 5 inch screen devices exist that run on mobile OSes and have the ability to make phone calls, such as the discontinued Dell Streak and the current Samsung Galaxy Note. Ergonomics arguments have been made that increasing screen sizes start to negatively impact usability.

Common resolutions for smartphone screens vary from 240×320 to 720×1280, with many flagship Android phones at 480×800 or 540×960, the iPhone 4/4S at 640×960 and Galaxy Nexus and HTC Rezound at 720×1280.

Application stores

The introduction of Apple's App Store for the iPhone and iPod Touch in July 2008 popularized manufacturer-hosted online distribution for third-party applications focused on a single platform. Before this, smartphone application distribution was largely dependent on third-party sources providing applications for multiple platforms, such as GetJar, Handango, Handmark, PocketGear, and others.

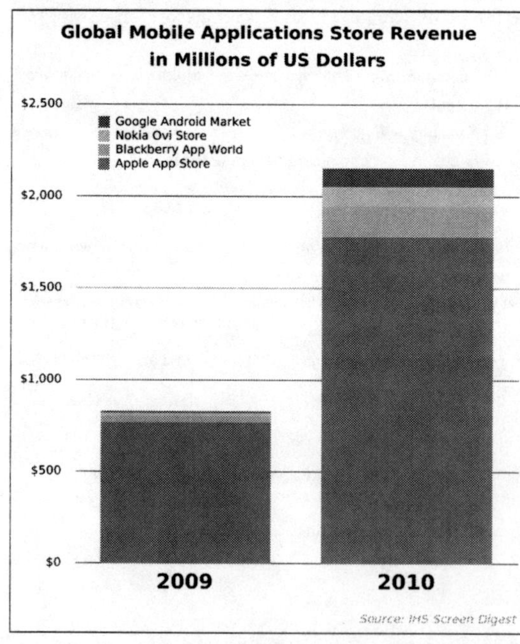

The iPhone's platform is officially restricted to installing apps through the App Store, through "B2B" deployment, and on an "Ad Hoc" basis on up to 100 iPhones.[164] Through jailbreaking it can install apps from other sources. Other platforms may allow application distribution through additional sources outside of their manufacturer-provided app stores, such as third-party app stores and downloads from individual websites.

Following the success of Apple's App Store other smartphone manufacturers quickly launched application stores of their own. Google launched the Android Market in October 2008. RIM launched its app store, BlackBerry App World, in April 2009. Nokia launched its Ovi Store in May 2009. Palm launched its Palm App Catalog for webOS in June 2009. Microsoft launched an application store for Windows Mobile called Windows Marketplace for Mobile in October 2009, and then a separate Windows

Phone Marketplace for Windows Phone in October 2010. Samsung launched Samsung Apps for its Bada based phones in June 2010. Amazon launched its Amazon Appstore for the Google Android operating system in March 2011.

Store	2009 (millions U.S.)	2010 (millions U.S.)[165]
Apple App Store	$769	$1782
Blackberry App World	$36	$165
Nokia Ovi Store	$13	$105
Google Android Market	$11	$102
Total	$828	$2155

The relatively high revenue of U.S. $1782 million in 2010 for Apple's App Store compared to competitor's stores[165] can be attributed to a combination of factors. In large part this can be attributed to having the largest number of apps available and the highest download volume of any mobile app store in 2010, but besides that only 28% of the apps in Apple's App Store were free apps, compared to over 57% in the Android Market. Similarly, Nokia's Ovi Store and the BlackBerry App World both had only 26% of their apps available for free, but both generated higher revenues than the Android Market despite having much lower download volumes.[166]

Malicious software attacks

As smartphone adoption goes up they have increasingly become subject to attacks by malicious software (malware).[167] [168]

Frequently this malware is distributed through application stores that have minimal or no review process for their content.[169] In some cases malware has been hidden in pirated versions of legitimate apps, which are then distributed through 3rd party app stores.[170] [171] Malware risk also comes from what's known as an "update attack," where a legitimate application is later changed to include a malware component, which users then install when they are notified that the app has been updated. Additionally, the ability to acquire software directly from links on the web results in a distribution vector called "malvertizing," where users are directed to click on links, such as on ads that look legitimate, which then open in the device's web browser and cause malware to be downloaded and installed automatically.[172]

Typical smartphone malware leverages platform vulnerabilities that allow it to gain root access on the device in the background. Using this access the malware installs additional software to target communications, location, or other personal identifying information. A common form of malware on mobile phones is the SMS trojan, which sends premium SMS messages, possibly while unknowingly running in the background of a legitimate application. These premium SMS messages run up charges on the owners phone bill which cannot be recovered.

In August 2010, Kaspersky Lab reported detection of the first malicious program for smartphones running on Google's Android operating system, named Trojan-SMS.AndroidOS.FakePlayer.a, an SMS trojan which had already infected a number of devices using that OS.[173] Over the spring of 2011 Android malware increased 76%, according to McAfee.[167] [174] A report from Juniper Global Threat Center notes that malware on the Android platform increased 400% from 2009 to the summer of 2010, and then saw a 472% increase between July and November 2011.[169] The Juniper report indicates that 55% of Android malware acts as spyware, and 44% are SMS trojans.

While there have been and continue to be potential security flaws in iOS,[175] as of at least August 2011 there were no known malware or spyware apps in Apple's App Store, according to security firm Lookout. There are however commercial spyware applications available, outside the App Store, for jailbroken iOS devices.[172] In June 2011 Symantec's 23-page report "A Window Into Mobile Device Security" characterized (non-jailbroken) devices running iOS as having "full protection" against malware attacks.[176]

Symbian and older versions Windows Mobile have had to contend with a degree of malware in the past, but as legacy systems it is believed that the people who previously targeted them have shifted their focus to Android.[169] There were also a few Palm OS viruses.

The only mobile platform other than Apple's iOS without reports of malware so far is HP's (formerly Palm's) webOS, but this may be explained by its relatively low adoption rate.[174]

The best way to reduce a device's vulnerability to malware attacks is to install the most recent versions of operating systems which include security patches. This can be complicated by long delays[177] in software updates for many devices which have had their software modified with custom "skins," services, or promotional on-deck apps by their manufacturer or mobile carrier.[172] In some cases a device may no longer be receiving updates from its manufacturer or carrier, leaving it vulnerable to exploits that have been patched in an OS version that's more recent than the device's last supported one.

Market share

Smartphone market share

For several years, demand for advanced mobile devices boasting powerful processors and graphics processing units, abundant storage (flash memory) for applications and media files, high-resolution screens with multi-touch capability, and open operating systems has outpaced the rest of the mobile phone market.[178]

According to an early 2010 study by ComScore, over 45.5 million people in the United States owned smartphones out of 234 million total subscribers.[179] Despite the large increase in smartphone sales in the last few years, smartphone shipments only made up 20% of total handset shipments as of the first half of 2010.[180]

According to Gartner in their report dated November 2010, total smartphone sales doubled in one year and now smartphones represent 19.3 percent of total mobile phone sales.[181] Smartphone sales increased in 2010 by 72.1 percent from the prior year, whereas sales for all mobile phones only increased by 32%.[182] [183]

According to an Olswang report in early 2011, the rate of smartphone adoption is accelerating: as of March 2011 22% of UK consumers had a smartphone, with this percentage rising to 31% amongst 24- to 35-year-olds.[184]

In March 2011, Berg Insight reported data that showed global smartphone shipments increased 74% from 2009 to 2010.[185]

A survey of mobile users in the United States by Nielsen in Q3, 2011 reports that smartphone ownership has reached 43% of all U.S. mobile subscribers, with the vast majority of users under the age of 44 owning one. In the 25-34 age range smartphone ownership is reported to be at 62%.[186] NPD Group reports that the share of handset sales that were smartphones in Q3, 2011 reached 59% for consumers 18 and over in the U.S.[187]

In profit share worldwide smartphones now far exceed the share of non-smartphones. According to a November 2011 research note from Canaccord Genuity, Apple Inc. holds 52% of the total mobile industry's operating profits, while only holding 4.2% of the global handset market. HTC and RIM similarly only make smartphones and their wordwide profit shares are at 9% and 7%, respectively. Samsung, in second place after Apple at 29%, makes both smartphones and feature phones and doesn't report a breakdown separating their profits between the two kinds of devices, but it can be intuited that a significant portion of that profit comes from their flagship smartphone devices.[188]

Up to the end of November 2011, camera-equipped smartphones took 27 percent of photos, a significant increase from 17 percent last year. Due to the fact that we carry smartphones with us all the time, smartphones have replaced some functions of Point-and-shoot cameras, except the cameras with big optical zoom such as 10x.[189]

Operating system market shares

2010 saw the rapid rise of the Google Android operating system from 4 percent of new deployments in 2009 to 33 percent at the beginning of 2011 making it share the top position with the since long dominating Symbian OS. The smaller rivals include US popular Blackberry OS, iOS, Samsung's recently introduced Bada, HP's heir of Palm webOS and the Microsoft Windows Phone OS which is now supported by Nokia.

Quantity market shares by Gartner (in one year) (new sales)	
Smartphone OS	**Percent**
Android 2009	3.9%
Android 2010	22.7%
Symbian 2009	46.9%
Symbian 2010	37.6%
RIM 2009	19.9%
RIM 2010	16.0%
iOS 2009	14.4%
iOS 2010	15.7%
Windows 2009	8.7%
Windows 2010	4.2%
Other 2009	6.1%
Other 2010	3.8%

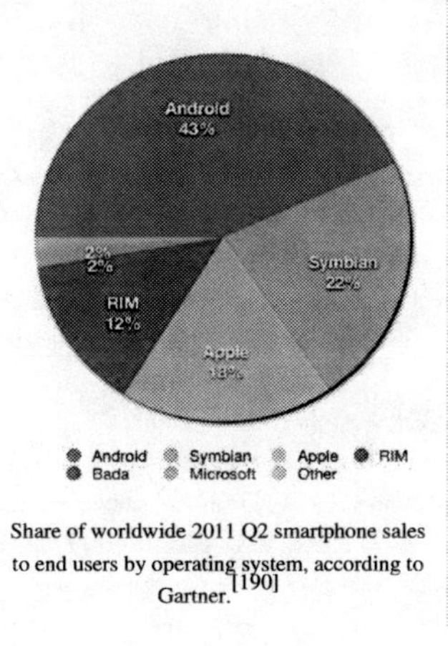

Share of worldwide 2011 Q2 smartphone sales to end users by operating system, according to Gartner.[190]

Over late 2009 and 2010 Android's smartphone operating system market share increased very rapidly.[181] In the fourth quarter of 2010, Android surpassed Symbian as the most common operating system in smartphones, with 32.9 million units sold versus 31.0 million. Android-equipped phones sold seven times more than in the prior year.[191] According to Canalys, Google's Android operating system, which is offered to phone makers for free, has raced to the top past operating systems by Nokia, Apple, RIM, and Microsoft. In Q1 2011 Google's Android market share was 35 percent, increasing significantly from 10 percent the previous year, while Nokia's Symbian dropped to 26 percent from 46 percent over the same time period.[192] In the UK, which currently has one of the highest penetrations of smartphones in the World, Android achieved 50% market share in October 2011.[193]

Historical market share

Figures in millions.

Year	Android (Google)	Blackberry (RIM)	IPhone (Apple)	Linux	Palm/WebOS (Palm/HP)	Symbian (Nokia)	Windows Mobile/Phone (Microsoft)
2007[194]		11.77	3.3	11.76	1.76	77.68	14.7
2008[195]		23.15	11.42	11.26	2.51	72.93	16.5
2009[196]	6.8	34.35	24.89	8.13	1.19	80.88	15.03
2010[197]	67.22	47.45	46.6			111.58	12.38

Enterprise share by operating system

In a worldwide study of 2,300 workers at 1,100 businesses by iPass it was reported that Apple's iPhones have displaced RIM's BlackBerry devices in enterprise adoption in 2011.[198] The share for iPhones increased to 45% from 31.1% in 2010, while the Blackberry share dropped to 32.2% from 34.5% in the previous year. Android phones also increased in share, to 21.3% from 11.3% in 2010, exceeding Symbian for the first time, which dropped to 7.4% from 12.4%. Windows Mobile and all other smartphone OSes also dropped in 2011 compared to 2010.[199]

Customer loyalty by operating system

According to a survey of more than 6,000 smartphone users through 2010 by mobile analytics firm Zokem, the top five loyalty scores for smartphone platforms are the iPhone at 73%, followed by Google's Android at 40%, Samsung's Bada at 33%, RIM's BlackBerry at 30%, and Symbian S60 at 23%. Windows Mobile and Palm follow at 10% each. Customer loyalty gauges the likelihood that the user of a smartphone platform whose contract has expired or who has broken or lost their phone will repurchase another one that uses that same platform.[200] [201]

Manufacturer market shares

From the launch of their Communicator model in 1996 until 2011 Nokia was dominant in the smartphone market, though has more recently been joined by other competitors in the market. Based on a report by Strategy Analytics, Samsung overtook Nokia in smartphone shipments with an estimated 27.8 million units shipped in Q3 2011[202] (Samsung does not publicly disclose the numbers of their smartphone shipments and sales).

Quantity market shares by Strategy Analytics (note that Gartner instead shows Nokia ahead on Apple sales based on Symbian sales only[203] (new sales)

Manufacturer	Percent
Apple Q2 2010	13.5%
Apple Q2 2011	18.5%
Samsung Q2 2010	5.0%
Samsung Q2 2011	17.5%
Nokia Q2 2010	38.1%
Nokia Q2 2011	15.2%
Others Q2 2010	43.4%
Others Q2 2011	48.9%

Market share among smartphone manufacturers does not resemble smartphone OS market share numbers due to the differences between the two major smartphone OS sales models: single manufacturer and licensed. Apple's iPhone, Nokia's Symbian, and RIM's BlackBerry smartphones are currently only available from single manufacturers.

Google's Android OS and Microsoft's mobile OSes are platforms that are licensed and used by a variety of manufacturers. As a result, manufacturers of smartphones using licensed OSes all split the total market share of that OS between them, while the total share for a single-manufacturer OS is held by that manufacturer alone.

Note that Nokia's Symbian OS was previously available from several manufacturers under a licensed model, then later predominantly only by Nokia itself more like a single manufacturer model.

Samsung smartphones use a diverse portfolio of operating systems, including their own Bada operating system along with Android and Windows Mobile.[204]

Apple surpassed Nokia worldwide by revenue and profit for the first time in Q2 2011 (though not in market share), with Apple's profit share of the total worldwide smartphone market increasing to 66.3% while Nokia reported a loss.[205]

Between Q2 2010 and Q2 2011 Nokia's worldwide Symbian smartphone sales dropped significantly from 38.1 percent to 15.2 percent, while Samsung smartphone sales increased significantly worldwide from 5% to 17.5%.[206] As of Q1 2011, Nokia had already announced plans to switch to Windows Phone.

Smartphone Customer Satisfaction by J.D. Power and Associates

Manufacturer	Score
Apple 2010	810
Apple 2011	838
HTC 2010	727
HTC 2011	801
Industry Average 2010	753
Industry Average 2011	788
Samsung 2010	724
Samsung 2011	777
Motorola 2010	N/A
Motorola 2011	775
RIM 2010	741
RIM 2011	762
LG 2010	N/A
LG 2011	760
HP/Palm 2010	712
HP/Palm 2011	733
Nokia 2010	720
Nokia 2011	721

Rankings are based on a possible top score of 1000

Nokia still remains the number one company in the worldwide mobile phone market with sales for Q2 2011 of 88.5 million when including feature phone platforms such as S40, compared with 16.7 million smartphones running Symbian.[207]

According to Nielsen in July 2011, in the United States Apple is the top smartphone manufacturer at 28% of the market, with RIM at 20%. Google Android has 39% of the U.S. market as a whole, but this is split between HTC at 14%, Motorola at 11%, Samsung at 8%, and other remaining manufacturers at 6%. HTC's total share of the U.S.

smartphone market actually ties RIM at 20%, since sales of their smartphones running Microsoft's mobile operating systems account for 6% of the total market. Samsung similarly gains 2% of overall U.S. market share due to their sales of Microsoft OS-based smartphones. In contrast to the worldwide market, Nokia's share of U.S. smartphone sales is very small, at only 2%.[208] [209] Nielsen's Q3, 2011 survey of mobile users maintains Apple as the top U.S. smartphone maker with a continued 28% of the market, with RIM dropping from 20% to 18%.[186] While Google Android increased in total operating system share from 39% to 43% of the U.S. market, it remains fragmented amongst many different manufacturers. Over the same quarter Microsoft managed a modest gain from 6% to 7% total U.S. smartphone OS share.

Checks with U.S. carriers by technology analyst firm Canaccord Genuity in April and August 2011 have found that Apple's iPhone 4 has consistently been the top selling device at AT&T and Verizon. In addition, the second most popular spot at AT&T has been maintained by the iPhone 3GS, which was originally released in 2009 (and has never been sold on Verizon). In August 2011 the most popular smartphones on Sprint and T-Mobile in the U.S. were the HTC EVO 3D 4G and HTC Sensation, respectively. The other second most popular smartphones were the Samsung Charge 4G on Verizon, the Motorola Photon 4G on Sprint, and the HTC myTouch 4G Slide on T-Mobile.[210] [211] NPD Group reported that in Q3, 2011 the overall top 5 smartphones by sales across all carriers in the U.S. were, in order: the iPhone 4, iPhone 3GS, HTC EVO 4G, Motorola Droid 3, and Samsung Intensity II.[187]

Currently the vast majority of smartphones are manufactured in China, Taiwan and Mexico, for companies based in the U.S. (Apple, HP, Motorola), South Korea (LG, Samsung), Canada (RIM), Finland (Nokia), Taiwan (HTC) and the U.K. (Sony Ericsson).

In Q3 2011, Samsung became the world's number one smartphone vendor, selling 24 million smartphones. Nokia's Symbian platform remained in second place with 19.5 million Symbian smartphones, and Apple fell to third place, with 17 million phones. Android had 52.5% of the market, with 60.5 million Android phones being sold. In the US, HTC became the largest smartphone vendor.[212]

Customer satisfaction by manufacturer

According to global marketing information services firm J.D. Power and Associates smartphones from Apple Inc. have been consistently[213] ranking highest in customer satisfaction,[214] [215] with a late 2011 score of 838 out of 1000. Based on the responses to their most recent survey of 6,898 smartphone users, Apple was followed in ranking by HTC (801), Samsung (777), Motorola (775), RIM (762), LG (760), Palm (733), and Nokia (721).[216] [217] [218] [219]

Open-source development

The open-source culture has penetrated the smartphone market in several ways. There have been attempts to open source both hardware and software of smartphones.

In February 2010 Nokia made Symbian open source. Thus, most commercial smartphones were based on open-source operating systems. These include those based on Linux, such as Google's Android, Nokia's Maemo, Hewlett-Packard's webOS, and those based on BSD, such as the Darwin-based Apple iOS. Maemo was later merged with Intel's project Moblin to form MeeGo.[220] [221]

Popular services

Location-based check-in services

According to a ComScore report released on May 12, 2011, nearly one in five smartphone users are tapping into check-in services like Foursquare and Gowalla. A total of 16.7 million mobile-phone subscribers used location-based services on their phones in March 2011.[222]

Second screen

The smartphones have introduced a new way of watching television.[223] The second screen is a consequence of the media multitasking which is exploding.[224]

See also

- Camera phone and videophone
- Comparison of smartphones
- List of digital distribution platforms for mobile devices
- Mobile broadband connectivity
- Mobile Internet device (MID) and personal digital assistant (PDA)
- Mobile operating system
- Second screen

References

[1] "Smartphone" (http://www.phonescoop.com/glossary/term.php?gid=131). *Phone Scoop*. . Retrieved 2011-12-15.

[2] "Feature Phone" (http://www.phonescoop.com/glossary/term.php?gid=310). *Phone Scoop*. . Retrieved 2011-12-15.

[3] Andrew Nusca (20 August 2009). "Smartphone vs. feature phone arms race heats up; which did you buy?" (http://www.zdnet.com/blog/gadgetreviews/smartphone-vs-feature-phone-arms-race-heats-up-which-did-you-buy/6836). ZDNet. . Retrieved 2011-12-15.

[4] "Smartphone definition from PC Magazine Encyclopedia" (http://www.pcmag.com/encyclopedia_term/0,2542,t=Smartphone&i=51537,00.asp). *PC Magazine*. . Retrieved 2011-12-15.

[5] Schneidawind, J: "Big Blue unveiling", *USA Today*, November 23, 1992, p. 2B.

[6] "Ericsson GS88 Preview" (http://pws.prserv.net/Eri_no_moto/GS88_Preview.htm). *Eri-no-moto*. . Retrieved 2011-12-15.

[7] "History" (http://www.stockholmsmartphone.org/history/). Stockholm Smartphone. . Retrieved 2011-12-15.

[8] "Ericsson GS88 box" (http://www.stockholmsmartphone.org/wp-content/uploads/penelope-box.jpg). . Retrieved 2011-12-15.

[9] "PDA Review: Ericsson R380 Smartphone" (http://www.geek.com/hwswrev/pda/ericr380/). *Geek.com*. . Retrieved 2011-12-15.

[10] "Symbian Device – The OS Evolution" (http://www.i-symbian.com/wp-content/uploads/2009/11/Symbian_Evolution.pdf) (PDF). *Independent Symbian Blog*. . Retrieved 2011-12-15.

[11] "Ericsson Introduces The New R380e" (http://www.mobilemag.com/2001/09/25/ericsson-introduces-the-new-r380e). *Mobile Magazine*. . Retrieved 2011-12-15.

[12] "Ericsson R380 World, Review & Rating" (http://www.pcmag.com/article2/0,2817,40827,00.asp). *PCMag.com*. . Retrieved 2011-12-15.

[13] *Popular Science, December 1999* (http://books.google.com/books?id=8qSgh_Q-YOkC). *Google Books*. . Retrieved 2011-12-15.

[14] "Ericsson R380 PDA & Phone" (http://www.cellular.co.za/ericsson_r380.htm). *CellularOnline*. . Retrieved 2011-12-15.

[15] "Sony Ericsson P800" (http://www.allaboutsymbian.com/features/item/Sony_Ericsson_P800.php). *All About Symbian*. . Retrieved 2011-12-15.

[16] "Nokia N8 smartphone official site" (http://www.nokia.co.uk/gb-en/products/phone/n8-00/). . Retrieved 2011-12-15.

[17] "Nokia N8 review - best camera phone ever" (http://www.mobileburn.com/review.jsp?Id=11093). Mobileburn.com. 2010-10-05. . Retrieved 2011-12-15.

[18] "Mobile Choice Award: Best Sat-Nav" (http://www.mobilechoiceuk.com/News/mobile+choice+award:+Best+Sat-Nav/5274). Mobile Choice. 22 October 2010. . Retrieved 2011-12-15.

[19] "Nokia, Microsoft in pact to rival Apple, Google - Technology & Science" (http://www.cbc.ca/news/technology/story/2011/02/11/nokia-microsoft-smart-phone-apple-google.html). Associated Press. CBC.ca. 2011-02-11. . Retrieved 2011-12-15.

[20] Velazco, Chris (26 October 2011). "Nokia Debuts Their First Windows Phones: The Lumia 800 and Lumia 710" (http://techcrunch.com/2011/10/26/nokia-debuts-lumia-710-and-lumia-800/). *TechCruch*. . Retrieved 26 October 2011.

[21] "Kyocera QCP 6035 Smartphone Review" (http://www.palminfocenter.com/view_story.asp?ID=1707). Palminfocenter.com. 2001-03-16. . Retrieved 2011-09-07.

[22] Segan, Sascha (2010-03-23). "Kyocera Launches First Smartphone In Years | News & Opinion" (http://www.pcmag.com/article2/0,2817,2361664,00.asp). PCmag.com. . Retrieved 2011-09-07.

[23] "Better Living through Software: Microsoft Advances for the Home Highlighted at Consumer Electronics Show 2002: Microsoft showcases new and improved products -- including "Freestyle," "Mira" and Ultimate TV at International CES 2002" (http://www.microsoft.com/presspass/features/2002/Jan02/01-08msces.mspx). Microsoft.com. 2002-01-08. . Retrieved 2011-09-07.

[24] Stephen H. Wildstrom (November 30, 2001). "Handspring's Breakthrough Hybrid" (http://www.businessweek.com/bwdaily/dnflash/nov2001/nf20011129_0157.htm). Businessweek.com. . Retrieved 2011-12-15.

[25] Kevin McLaughlin (December 17, 2009). "BlackBerry Users Call For RIM To Rethink Service" (http://www.crn.com/news/client-devices/222002587/blackberry-users-call-for-rim-to-rethink-service.htm). CRN.com. . Retrieved 2011-12-15.

[26] "iPhone in depth: the Ars review" (http://arstechnica.com/apple/reviews/2007/07/iphone-review.ars/6). *ArsTechnica*. Condé Nast. 9 July 2007. p. 6. . Retrieved 3 August 2010.

[27] "iPhone to Support Third-Party Web 2.0 Applications" (http://www.apple.com/pr/library/2007/06/11iphone.html). *Press Release*. Apple Inc.. June 11, 2007. . Retrieved December 15, 2008.

[28] "The iPhone is not a smartphone" (http://www.engadget.com/2007/01/09/the-iphone-is-not-a-smartphone/). Engadget.com. 9 January 2007. . Retrieved 11 July 2010.

[29] "Smartphones Can Replace These Everyday Items" (http://simpleorganizedlife.com/smartphones-can-replace-these-everyday-items/). Simple. Organized. Life.. October 10, 2011. . Retrieved 2011-12-15.

[30] "iPhone 3G on Sale Tomorrow" (http://www.apple.com/pr/library/2008/07/10iphone.html). *Press Release*. Apple Inc.. 2008-07-10. . Retrieved 2009-01-17.

[31] "iPhone App Store Downloads Top 10 Million in First Weekend" (http://www.apple.com/pr/library/2008/07/14iPhone-App-Store-Downloads-Top-10-Million-in-First-Weekend.html). *Press Release*. Apple Inc.. July 14, 2008. . Retrieved 2011-12-15.

[32] "Apple's App Store Downloads Top 1.5 Billion in First Year" (http://www.apple.com/pr/library/2009/07/14Apples-App-Store-Downloads-Top-1-5-Billion-in-First-Year.html). *Press Release*. Apple Inc.. July 14, 2009. . Retrieved 2011-12-15.

[33] "Apple's App Store Downloads Top 15 Billion" (http://www.apple.com/pr/library/2011/07/07Apples-App-Store-Downloads-Top-15-Billion.html). *Press Release*. Apple Inc.. July 7, 2011. . Retrieved 2011-12-15.

[34] "Apple now accepting iOS 4 apps, multitasking ahoy" (http://www.engadget.com/2010/06/11/apple-now-accepting-ios-4-apps-multitasking-ahoy/). Engadget.com. 11 June 2010. . Retrieved 2011-12-15.

[35] "Apple Presents iPhone 4" (http://www.apple.com/pr/library/2010/06/07Apple-Presents-iPhone-4.html). *Press Release*. Apple Inc.. 2010-06-07. . Retrieved 2011-07-05.

[36] "Liveblog: The Verizon iPhone" (http://voices.washingtonpost.com/fasterforward/2011/01/liveblog_the_verizon_iphone.html). *The Washington Post*. .

[37] Memmott, Mark (2011-01-11). "It's Official: Verizon Has The iPhone 4 : The Two-Way" (http://www.npr.org/blogs/thetwo-way/2011/01/11/132833078/its-official-verizon-has-iphone-4). NPR. . Retrieved 2011-09-07.

[38] Raice, Shayndi (January 12, 2011). "Verizon Unwraps iPhone" (http://online.wsj.com/article/SB10001424052748703791904576075681886276172.html?mod=googlenews_wsj). *The Wall Street Journal*. .

[39] Glenn Fleishman (2011-02-22). "Using the Personal Hotspot on your Verizon iPhone" (http://www.macworld.com/article/158058/2011/02/personal_hotspot_verizon.html). Macworld.com. . Retrieved 2011-03-12.

[40] "iOS 4.3 Software Update" (http://www.apple.com/ios/). Apple Inc.. . Retrieved 2011-03-12.

[41] Dan Moren (2011-03-11). "Hands on with iOS 4.3" (http://www.macworld.com/article/158483/2011/03/firstlook_43.html). Macworld.com. . Retrieved 2011-03-12.

[42] "Apple Launches iPhone 4S, iOS 5 & iCloud" (http://www.apple.com/pr/library/2011/10/04Apple-Launches-iPhone-4S-iOS-5-iCloud.html). Apple Inc.. October 4, 2011. . Retrieved 2011-12-15.

[43] "iPhone 4S Pre-Orders Top One Million in First 24 Hours" (http://www.apple.com/pr/library/2011/10/10iPhone-4S-Pre-Orders-Top-One-Million-in-First-24-Hours.html). Apple. . Retrieved 10 October 2011.

[44] Anderson, Ash. "iPhone 4S Sells 1 Million in Under 24 Hours" (http://www.keynoodle.com/iphone-4s-sells-1-million-in-under-24-hours/). *KeyNoodle*. . Retrieved 2011-12-15.

[45] "Apple's fall from grace" (http://www.apple.com/pr/library/2011/10/04Apple-Launches-iPhone-4S-iOS-5-iCloud.html). BGR.com. October 5, 2011. . Retrieved 2011-12-15.

[46] Michael E. Cohen (October 13, 2011). "iPhone 4S: A Very Palpable Hit" (http://tidbits.com/article/12554). tidbits.com. . Retrieved 2011-12-15.

[47] David Pogue (October 11, 2011). "New iPhone Conceals Sheer Magic" (http://www.nytimes.com/2011/10/12/technology/personaltech/iphone-4s-conceals-sheer-magic-pogue.html?_r=2&pagewanted=all). The New York Times. . Retrieved 2011-12-15.

[48] "Press Info - Apple to Launch iCloud on October 12" (http://www.apple.com/pr/library/2011/10/04Apple-to-Launch-iCloud-on-October-12.html). Apple. 2011-10-04. . Retrieved 2012-01-05.

[49] "Press Info - New Version of iOS Includes Notification Center, iMessage, Newsstand, Twitter Integration Among 200 New Features" (http://www.apple.com/pr/library/2011/06/06New-Version-of-iOS-Includes-Notification-Center-iMessage-Newsstand-Twitter-Integration-Among-200-New-Features.html). Apple.

2011-06-06. . Retrieved 2012-01-05.

[50] Michael B. Farrell (November 12, 2011). "No cash, card? No problem: Paying by smartphone is an emerging trend" (http://articles.boston.com/2011-11-12/business/30391540_1_smartphone-mobile-commerce-card-readers). boston.com. . Retrieved 2011-12-15.

[51] 13 November 2011. (http://cellphones.about.com/od/smartphonebasics/a/what_is+_smart.htm)

[52] "Alliance Members" (http://www.openhandsetalliance.com/oha_members.html). *Open Handset Alliance*. . Retrieved 16 January 2011.

[53] "Weighing Nexus over iPhone – a practical review for the everyday user!" (http://techietrick.blogspot.com/2010/01/weighing-nexus-over-iphone-practical.html). Techietrick.blogspot.com. 2010-01-06. . Retrieved 2011-09-07.

[54] "Nexus One gets a software update, enables multitouch (updated with video!)" (http://www.engadget.com/2010/02/02/nexus-one-gets-a-software-update-enables-multitouch). Engadget.com. . Retrieved 2011-09-07.

[55] Tarmo Virki (2011-02-14). "Sony takes gaming console war to phones" (http://in.reuters.com/article/2011/02/13/idINIndia-54865020110213). Reuters. .

[56] "HTC EVO 3D" (http://web.archive.org/web/20110511105547/http://www.htc.com/www/product/evo3d/overview.html). htc.com. Archived from the original (http://www.htc.com/www/product/evo3d/overview.html) on 2011-05-11. . Retrieved 2011-12-15.

[57] Savov, Vlad. "HTC Evo 3D Launches June 24th" (http://www.engadget.com/2011/06/06/htc-evo-3d-launches-on-june-24th-for-200-joined-by-evo-view-4g/). Engadget.com. . Retrieved 7 June 2011.

[58] Ed Hansberry (11 November 2009). "Samsung Bailing on Windows Mobile" (http://www.informationweek.com/blog/main/archives/2009/11/samsung_bailing.html). *InformationWeek*. .

[59] "Samsung to Discard Windows Phone" (http://www.telecomskorea.com/market-8281.html). *Telecoms Korea*. 9 November 2009. .

[60] "Samsung Wave, first Bada smartphone hits the market" (http://www.bada.com/samsung-wave-first-bada-smartphone-hits-the-market/). *Bada*. 24 May 2010. . Retrieved 3 February 2011.

[61] "BadaWave" (http://badawave.com/). BadaWave. . Retrieved 2012-01-05.

[62] "Samsung Waves away a million" (http://www.theinquirer.net/inquirer/news/1722287/samsung-waves-away-million). The Inquirer. 13 July 2010. .

[63] "Android increases smart phone market leadership with 35% share" (http://www.canalys.com/newsroom/android-increases-smart-phone-market-leadership-35-share). Canalys.com. 2011-05-04. . Retrieved 2011-09-07.

[64] "Samsung Bada shipments up 355% to 4.5 million units in Q2 2011 | asymco news | PG.Biz" (http://www.pocketgamer.co.uk/r/PG.Biz/asymco+news/news.asp?c=32049). Pocket Gamer. . Retrieved 2011-09-07.

[65] Florian Mueller. "Apple vs Android 10.12.02" (http://www.scribd.com/doc/44759893/Apple-vs-Android-10-12-02). . Retrieved 2011-09-08.

[66] Florian Mueller. "NokiaVsApple_11.03.31.100" (http://www.scribd.com/doc/52195210/NokiaVsApple-11-03-31-100). . Retrieved 2011-09-08.

[67] Florian Mueller. "Microsoft vs Motorola 11.04.09" (http://www.scribd.com/doc/52754079/Microsoft-vs-Motorola-11-04-09). . Retrieved 2011-09-08.

[68] Florian Mueller. "AppleVsSamsung_11.07.05" (http://www.scribd.com/doc/59474521/AppleVsSamsung-11-07-05). . Retrieved 2011-09-08.

[69] Florian Mueller. "AppleVsHTCandS3_11.07.29" (http://www.scribd.com/doc/61387056/AppleVsHTCandS3-11-07-29). . Retrieved 2011-09-08.

[70] "Nokia sues Apple over iPhone's use of patented wireless standards" (http://www.appleinsider.com/articles/09/10/22/nokia_sues_apple_over_iphones_use_of_patented_wireless_standards.html). Appleinsider.com. 2009-10-22. . Retrieved 2012-01-05.

[71] "Nokia vs. Apple: the in-depth analysis" (http://www.engadget.com/2009/10/29/nokia-vs-apple-the-in-depth-analysis/). Engadget.com. . Retrieved 2012-01-05.

[72] "Apple countersues Nokia for infringing 13 patents" (http://www.engadget.com/2009/12/11/apple-countersues-nokia-for-infringing-13-patents/). Engadget.com. . Retrieved 2012-01-05.

[73] Foresman, Chris (2010-01-04). "Nokia adds additional lawsuit in patent catfight with Apple" (http://arstechnica.com/apple/news/2010/01/nokia-adds-additional-lawsuit-in-patent-catfight-with-apple.ars). Arstechnica.com. . Retrieved 2012-01-05.

[74] Foresman, Chris (2009-12-29). "Nokia hurls new salvo in spat with Apple, complains to ITC" (http://arstechnica.com/apple/news/2009/12/nokia-hurls-new-salvo-in-spat-with-apple-complains-to-itc.ars). Arstechnica.com. . Retrieved 2012-01-05.

[75] Decker, Susan (2010-01-16). "Apple Files New Trade Complaint With Against Nokia" (http://www.bloomberg.com/apps/news?pid=newsarchive&sid=ao_5HVbD_IRM). Bloomberg.com. . Retrieved 2012-01-05.

[76] Cheng, Jacqui (2010-01-18). "Apple wants Nokia's US imports blocked" (http://arstechnica.com/apple/news/2010/01/apple-continues-nokia-patent-spat-with-its-own-itc-complaint.ars). Arstechnica.com. . Retrieved 2012-01-05.

[77] "Apple vs HTC: a patent breakdown" (http://www.engadget.com/2010/03/02/apple-vs-htc-a-patent-breakdown/). Engadget.com. 2010-03-02. . Retrieved 2012-01-05.

[78] "Google backs HTC in what could be 'long and bloody battle' with Apple" (http://www.appleinsider.com/articles/10/03/03/google_backs_htc_in_what_could_be_long_and_bloody_battle_with_apple.html). Appleinsider.com. 2010-03-03. . Retrieved 2012-01-05.

[79] Bilton, Nick (2010-03-02). "What Apple vs. HTC Could Mean" (http://bits.blogs.nytimes.com/2010/03/02/what-apple-vs-htc-could-mean/). Bits.blogs.nytimes.com. . Retrieved 2012-01-05.

[80] "HTC says it uses own technology, not Apple's" (http://www.macworld.com/article/146819/2010/03/htc_defense.html). Macworld.com. . Retrieved 2012-01-05.

[81] Foresman, Chris (2010-03-09). "HTC lawsuit came after warning by Apple to handset makers" (http://arstechnica.com/apple/news/2010/03/htc-lawsuit-came-after-warning-by-apple-to-handset-makers.ars). Arstechnica.com. . Retrieved 2012-01-05.
[82] "Microsoft Announces Patent Agreement With HTC" (http://www.microsoft.com/Presspass/press/2010/apr10/04-27MSHTCPR.mspx). Microsoft.com. 2010-04-27. . Retrieved 2012-01-05.
[83] "HTC licenses Microsoft's mobile patents for Android phones – First Take" (http://gartenberg.wordpress.com/2010/04/28/htc-licenses-microsofts-mobile-patents-for-android-phones-first-take/). Gartenberg.wordpress.com. 2010-04-28. . Retrieved 2012-01-05.
[84] Johnston, Casey (2010-05-07). "Nokia applies patent thumbscrews to Apple's iPad" (http://arstechnica.com/apple/news/2010/05/nokia-applies-patent-thumbscrews-to-apples-ipad.ars). Arstechnica.com. . Retrieved 2012-01-05.
[85] "HTC countersues Apple, claims infringement of five patents" (http://www.appleinsider.com/articles/10/05/12/htc_countersues_apple_claims_infringement_of_five_patents.html). Appleinsider.com. 2010-05-12. . Retrieved 2012-01-05.
[86] "S3 Graphics Files New 337 Complaint Regarding Certain Electronic Devices With Image Processing Systems" (http://www.itcblog.com/20100602/s3-graphics-files-new-337-complaint-regarding-certain-electronic-devices-with-image-processing-systems/). Itcblog.com. 2010-06-02. . Retrieved 2012-01-05.
[87] August 12, 2010 (2010-08-12). "Oracle sues Google over Android" (http://venturebeat.com/2010/08/12/oracle-sues-google-over-android/). Venturebeat.com. . Retrieved 2012-01-05.
[88] Protalinski, Emil (2010-10-02). "Microsoft sues Motorola, citing Android patent infringement" (http://arstechnica.com/microsoft/news/2010/10/microsoft-sues-motorola-citing-android-patent-infringement.ars). Arstechnica.com. . Retrieved 2012-01-05.
[89] "Microsoft Files Patent Infringement Action Against Motorola" (http://www.microsoft.com/presspass/press/2010/oct10/10-01statement.mspx). Microsoft.com. 2010-10-01. . Retrieved 2012-01-05.
[90] Foresman, Chris (2010-10-06). "Motorola asks ITC, two federal courts to throw book at Apple" (http://arstechnica.com/apple/news/2010/10/motorola-asks-itc-two-federal-courts-to-throw-book-at-apple.ars). Arstechnica.com. . Retrieved 2012-01-05.
[91] "Motorola Moves Offensively with a New Lawsuit against Apple" (http://www.patentlyapple.com/patently-apple/2010/10/motorola-moves-offensively-with-a-new-lawsuit-against-apple.html). Patentlyapple.com. 2010-10-15. . Retrieved 2012-01-05.
[92] Motorola seeks to invalidate Apple phone patents (http://arstechnica.com/apple/news/2010/10/motorola-horns-in-on-apple-vs-htc-files-suit-against-apple.ars).
[93] Apple Files Lawsuit against Motorola to Defend Multi-Touch (http://www.patentlyapple.com/patently-apple/2010/10/apple-files-lawsuit-against-motorola-to-defend-multi-touch.html).
[94] Apple Patent Case Against Motorola to Get Trade Agency Review (http://www.businessweek.com/news/2010-11-23/apple-patent-case-against-motorola-to-get-trade-agency-review.html).
[95] Apple vs. Motorola: now 42 patents-in-suit (24 Apple and 18 Motorola patents) (http://fosspatents.blogspot.com/2010/12/apple-vs-motorola-now-42-patents-in.html).
[96] Microsoft sues Barnes & Noble over Android in Nook (http://www.geekwire.com/2011/breaking-microsoft-sues-barnes-noble-android).
[97] Nokia files second ITC complaint against Apple (http://press.nokia.com/2011/03/29/nokia-files-second-itc-complaint-against-apple/).
[98] FOSS Patents: Escalation: Nokia files new ITC complaint against Apple (plus a corresponding federal lawsuit) (http://fosspatents.blogspot.com/2011/03/escalation-nokia-files-new-itc.html).
[99] "Apple to Samsung: Stop stealing ideas" (http://www.cnn.com/2011/TECH/innovation/04/19/apple.samsung.lawsuit.wired/). *CNN.com*. . Retrieved 19 April 2011.
[100] Apple sues Samsung: a complete lawsuit analysis (http://thisismynext.com/2011/04/19/apple-sues-samsung-analysis/).
[101] Samsung Countersues Apple for Patent Infringement (http://www.pcmag.com/article2/0,2817,2383964,00.asp).
[102] Samsung sues Apple for infringing 10 patents: a closer look (http://thisismynext.com/2011/04/29/samsung-sues-apple-infringing-10-patents-closer/).
[103] Samsung Must Show Cell Phones to Apple (http://www.courthousenews.com/2011/05/19/36708.htm).
[104] Judge Orders Samsung to Give Apple Unreleased Tablets, Smartphones (http://www.pcmag.com/article2/0,2817,2385861,00.asp).
[105] Samsung's lawyers demand to see the iPhone 5 and iPad 3 (http://thisismynext.com/2011/05/28/samsung-apple-iphone-5-ipad-3/).
[106] Nokia Wins Apple Patent-License Deal Cash, Settles Lawsuits (http://www.bloomberg.com/news/2011-06-14/nokia-apple-payments-to-nokia-settle-all-litigation.html).
[107] Apple Settles With Nokia In Patent Lawsuit (http://www.huffingtonpost.com/2011/06/14/apple-nokia-patent-lawsuit-settlement_n_876499.html).
[108] Staff Reporter, "Apple alleges Samsung of 'slavishly copying' its technology; questions new Galaxy Tab 10.1" (http://www.ibtimes.com/articles/165405/20110619/apple-lawsuit-upgrade-samsung-galaxy-tab-10-1-infringement-intellectual-property-rights-ipad-2.htm), *International Business Times*, 19 June 2011.
[109] Shocker: Samsung not allowed to see iPhone 5 and iPad 3 (http://thisismynext.com/2011/06/22/shocker-samsung-allowed-iphone-5-ipad-3/).
[110] "Microsoft and General Dynamics Itronix Sign Patent Agreement: Agreement will cover General Dynamics Itronix devices running the Android platform" (http://www.microsoft.com/Presspass/press/2011/jun11/06-27ItronixPR.mspx). Microsoft.com. 2011-06-27. . Retrieved 2012-01-05.
[111] FOSS Patents: Android device makers sign up as Microsoft patent licensees -- an overview of the situation (http://fosspatents.blogspot.com/2011/06/android-device-makers-sign-up-as.html).

[112] "Microsoft and Velocity Micro, Inc., Sign Patent Agreement Covering Android-Based Devices: Agreement provides broad coverage of Microsoft's patent portfolio" (http://www.microsoft.com/Presspass/press/2011/jun11/06-29VelocityMicroPR.mspx). Microsoft.com. 2011-06-29. . Retrieved 2012-01-05.

[113] Dealtalk: Google bid "pi" for Nortel patents and lost (http://www.reuters.com/article/2011/07/02/us-dealtalk-nortel-google-idUSTRE76104L20110702).

[114] Nortel Announces the Winning Bidder of Its Patent Portfolio for a Purchase Price of US$4.5 Billion (http://www.marketwatch.com/story/nortel-announces-the-winning-bidder-of-its-patent-portfolio-for-a-purchase-price-of-us45-billion-2011-06-30?reflink=MW_news_stmp).

[115] Who Won The 6,000+ Nortel Patents? Apple, RIM, Microsoft — Everyone But Google (http://techcrunch.com/2011/07/01/apple-microsoft-rim-google-nortel-patents/).

[116] "Microsoft and Onkyo Corp. Sign Patent Agreement Covering Android-Based Tablets: Agreement provides broad coverage of Microsoft's patent portfolio" (http://www.microsoft.com/Presspass/press/2011/jun11/06-30OnkyoPR.mspx?rss_fdn=Custom). Microsoft.com. . Retrieved 2012-01-05.

[117] Apple files motion for preliminary injunction in the U.S. against four Samsung products: Infuse 4G, Galaxy S 4G, Droid Charge, Galaxy Tab 10.1 (http://fosspatents.blogspot.com/2011/07/apple-files-motion-for-preliminary.html).

[118] ITC ruling mixed in S3 Graphics v. Apple (http://news.cnet.com/8301-27076_3-20076219-248/itc-ruling-mixed-in-s3-graphics-v-apple/?tag=mncol;txt).

[119] "Microsoft and Wistron Sign Patent Agreement: Agreement will cover Wistron's Android tablets, smartphones and e-readers" (http://www.microsoft.com/Presspass/press/2011/jul11/07-05WistronPR.mspx). Microsoft.com. 2011-07-05. . Retrieved 2012-01-05.

[120] Microsoft patent division taking cash from at least 5 Android vendors (http://www.networkworld.com/news/2011/070511-microsoft-patent-android.html).

[121] DailyTech - VIA, WTI Sell Stakes in S3 Graphics to HTC (http://www.dailytech.com/VIA+WTI+Sell+Stakes+in+S3+Graphics+to+HTC/article22078.htm).

[122] HTC to Acquire Chip Designer S3 Graphics, Securing Patents in Apple Fight - Bloomberg (http://www.bloomberg.com/news/2011-07-06/htc-to-buy-s3-graphics-securing-patents-in-apple-fight-1-.html).

[123] HTC facing backlash for S3 acquisition (http://news.cnet.com/8301-1035_3-20077720-94/htc-facing-backlash-for-s3-acquisition/?tag=mncol;txt).

[124] Microsoft wants Samsung to pay smartphone license (http://www.reuters.com/article/2011/07/06/us-samsung-microsoft-idUSTRE7651DB20110706).

[125] HTC's Flyer Tablet, Droid Accused of Infringing Apple Patents - Bloomberg (http://www.bloomberg.com/news/2011-07-11/apple-files-new-trade-complaint-against-htc-over-devices.html?cmpid=yhoo).

[126] FOSS Patents: Apple files second ITC complaint against HTC: better luck next time? (http://fosspatents.blogspot.com/2011/07/apple-files-second-itc-complaint.html).

[127] Google Acquires Over 1,000 IBM Patents in July (http://www.seobythesea.com/2011/07/google-acquires-ibm-patents-in-july/).

[128] Google's New Patents from IBM (http://www.seobythesea.com/2011/07/googles-new-patents-from-ibm/).

[129] Bad News for Android: ITC Rules HTC Violated Two Apple Patents - John Paczkowski - News - AllThingsD (http://allthingsd.com/20110715/itc-rules-htc-violated-two-apple-patents/?refcat=news).

[130] Apple Lawsuit Puts Samsung Tablet Sales in Australia on Hold (http://www.bloomberg.com/news/2011-08-01/apple-seeks-to-block-samsung-from-selling-tablet-in-australia.html).

[131] Samsung's official comment on the Australian Galaxy Tab 10.1 situation is extremely weak (http://fosspatents.blogspot.com/2011/08/samsungs-official-comment-on-australian.html).

[132] Tsukayama, Hayley (10 August 2011). "Samsung Galaxy Tab 10.1 imports banned in most of Europe" (http://www.washingtonpost.com/blogs/faster-forward/post/samsung-galaxy-tab-101-imports-banned-in-most-of-europe/2011/08/10/gIQAmnVr6I_blog.html). *The Washington Post*. . Retrieved 11 August 2011.

[133] Preliminary injunction granted by German court: Apple blocks Samsung Galaxy Tab 10.1 in the entire European Union except for the Netherlands (http://fosspatents.blogspot.com/2011/08/preliminary-injunction-granted-by.html).

[134] "Microsoft files to ban imports of Motorola smartphones" (http://androidandme.com/2011/08/news/microsoft-files-to-ban-imports-of-motorola-smartphones). Android and Me. . Retrieved 2011-12-10.

[135] Womack, Brian (2011-08-22). "Motorola Value Found in 18 Patents Used Against Apple: Tech" (http://www.bloomberg.com/news/2011-08-22/motorola-s-value-for-google-found-in-18-patents-used-against-apple-tech.html). Bloomberg. . Retrieved 2011-12-10.

[136] "Samsung Galaxy tablet ban lifted in most of Europe" (http://uk.reuters.com/article/2011/08/16/tech-us-samsung-apple-ban-idUKTRE77F45420110816). *Reuters*. . Retrieved 16 August 2011.

[137] "Samsung Galaxy Tab ban is on hold" (http://www.bbc.co.uk/news/business-14548895). *BBC*. . Retrieved 16 August 2011.

[138] Google and IBM do it again: Google Acquires over 1,000 Patents from IBM in August (http://www.seobythesea.com/2011/09/google-ibm-patents-august/#more-6675).

[139] Microsoft Asks For An Import Ban On Motorola Smartphones (http://techcrunch.com/2011/08/23/microsoft-asks-for-an-import-ban-on-motorola-smartphones/).

[140] "Microsoft files to ban imports of Motorola smartphones" (http://androidandme.com/2011/08/news/microsoft-files-to-ban-imports-of-motorola-smartphones/). Android and Me. . Retrieved 2011-12-10.

[141] Euro ban for Samsung Galaxy phone (http://www.bbc.co.uk/news/technology-14652482).
[142] Samsung Puts Galaxy 10.1 Tablet on Hold as Apple Wins German Court Order (http://www.bloomberg.com/news/2011-09-04/apple-wins-german-injunction-on-new-galaxy-tab.html).
[143] FOSS Patents: Apple to ITC: Andy Rubin got inspiration for Android framework while working at Apple, hence infringes an Apple API patent (http://fosspatents.blogspot.com/2011/09/apple-to-itc-andy-rubin-got-inspiration.html).
[144] Milford, Phil (2011-09-08). "HTC Sues Apple Using Google Patents Bought Last Week as Battle Escalates" (http://www.bloomberg.com/news/2011-09-07/htc-sues-apple-alleging-infringement-of-four-u-s-patents.html). Bloomberg. . Retrieved 2011-12-10.
[145] Google Hands HTC Patents to Use Against Apple in Smartphone Wars (http://www.businessweek.com/news/2011-09-08/google-hands-htc-patents-to-use-against-apple-in-smartphone-wars.html).
[146] Patel, Nilay (2011-09-07). "HTC sues Apple for patent infringement… using patents purchased by Google" (http://thisismynext.com/2011/09/07/htc-sues-apple-patent-infringement-patents-purchased-google/). Thisismynext.com. . Retrieved 2012-01-05.
[147] Google gets its hands dirty - Apple 2.0 - Fortune Tech (http://tech.fortune.cnn.com/2011/09/08/google-gets-its-hands-dirty/).
[148] "Microsoft and Acer Sign Patent License Agreement: Agreement will cover Acer's Android tablets and smartphones" (http://www.microsoft.com/Presspass/press/2011/sep11/09-08AcerPR.mspx). Microsoft.com. 2011-09-08. . Retrieved 2012-01-05.
[149] "Microsoft and ViewSonic Sign Patent Agreement: Agreement will cover ViewSonic's Android Tablets and smartphones" (http://www.microsoft.com/Presspass/press/2011/sep11/09-08ViewSonicPR.mspx). Microsoft.com. 2011-09-08. . Retrieved 2012-01-05.
[150] "Microsoft ropes two more OEMs into Android patent deal" (http://www.zdnet.com/blog/hardware/microsoft-ropes-two-more-oems-into-android-patent-deal/14622). Zdnet.com. . Retrieved 2012-01-05.
[151] "Microsoft signs Android patent agreements with Acer, ViewSonic" (http://seattletimes.nwsource.com/html/microsoftpri0/2016144500_microsoft_signs_android_patent_agreements_with_ace.html). Seattletimes.nwsource.com. 2011-09-08. . Retrieved 2012-01-05.
[152] Matussek, Karin (2011-09-09). "Apple Wins German Ban on Samsung Tablet" (http://www.bloomberg.com/news/2011-09-09/apple-wins-ruling-for-german-samsung-galaxy-tablet-10-1-ban.html). Bloomberg.com. . Retrieved 2012-01-05.
[153] "Latest Samsung lawsuit targets Apple's iPhone, iPad in France" (http://www.appleinsider.com/articles/11/09/13/latest_samsung_lawsuit_targets_apples_iphone_ipad_in_france.html). AppleInsider. 2011-09-13. . Retrieved 2012-01-05.
[154] Meyer, David (2011-09-14). "Apple sues Samsung in the UK over Android | ZDNet UK" (http://www.zdnet.co.uk/blogs/communication-breakdown-10000030/apple-sues-samsung-in-the-uk-over-android-10024342/). Zdnet.co.uk. . Retrieved 2012-01-05.
[155] "USPTO Assignments on the Web" (http://assignments.uspto.gov/assignments/q?db=pat&reel=026894&frame=0001). Assignments.uspto.gov. . Retrieved 2012-01-05.
[156] "Samsung fires back at Apple in Australia with countersuit against iPhone, iPad" (http://www.appleinsider.com/articles/11/09/16/samsung_files_patent_case_against_apple_in_australia_over_iphone_ipad.html). Appleinsider.com. . Retrieved 2012-01-05.
[157] "Microsoft and Samsung Broaden Smartphone Partnership: Agreements mark new initiatives to promote Windows Phone and share intellectual property" (http://www.microsoft.com/Presspass/press/2011/sep11/09-28SamsungPR.mspx). Microsoft.com. 2011-09-28. . Retrieved 2012-01-05.
[158] Our Licensing Deal with Samsung: How IP Drives Innovation and Collaboration - Microsoft on the Issues - Site Home - TechNet Blogs (http://blogs.technet.com/b/microsoft_on_the_issues/archive/2011/09/28/our-licensing-deal-with-samsung-how-ip-drives-innovation-and-collaboration.aspx).
[159] Florian Mueller (2011-09-28). "FOSS Patents: Samsung takes Android patent license from Microsoft rather than wait for Motorola" (http://fosspatents.blogspot.com/2011/09/samsung-takes-android-patent-license.html). Fosspatents.blogspot.com. . Retrieved 2012-01-05.
[160] "Apple Wins Injunction Blocking Sale of Galaxy Tab 10.1 in Australia" (http://www.macrumors.com/2011/10/12/apple-wins-injunction-blocking-sale-of-galaxy-tab-10-1-in-australia/). Mac Rumors. 2011-10-12. . Retrieved 2012-01-05.
[161] "Microsoft and Quanta Computer Sign Patent Agreement Covering Android and Chrome-Based Devices - Redmond, Wash., Oct. 13, 2011 /PRNewswire/" (http://www.prnewswire.com/news-releases/microsoft-and-quanta-computer-sign-patent-agreement-covering-android-and-chrome-based-devices-131793888.html). Washington: Prnewswire.com. . Retrieved 2012-01-05.
[162] Microsoft Inks Another Android Patent Deal, This Time With Quanta | TechCrunch (http://techcrunch.com/2011/10/13/microsoft-inks-another-android-patent-deal-this-time-with-quanta/).
[163] Levine, Dan. "U.S. judge says Samsung tablets infringe Apple patents" (http://www.reuters.com/article/2011/10/13/us-apple-samsung-lawsuit-idUSTRE79C79C20111013?feedType=RSS&feedName=businessNews&utm_source=dlvr.it&utm_medium=twitter&dlvrit=56943). Reuters.com. . Retrieved 2012-01-05.
[164] Apple Inc.. "Distribute your App - iOS Developer Program - Apple Developer" (http://developer.apple.com/programs/ios/distribute.html). Developer.apple.com. . Retrieved 2012-01-05.
[165] "Apple's rivals battle for iOS scraps as app market sales grow to $2.2 billion" (http://www.appleinsider.com/articles/11/02/18/rim_nokia_and_googles_android_battle_for_apples_ios_scraps_as_app_market_sales_grow_to_2_2_billion.html). Appleinsider.com. 2011-02-18. . Retrieved 2012-01-05.
[166] "Google Android has double the number of free apps than Apple's App Store" (http://techcrunch.com/2010/07/05/distimo-june-2010/). Distimo. 15 July 2009. . Retrieved 15 July 2009.
[167] "McAfee: Android malware surges 76%, iPhone untouched" (http://www.electronista.com/articles/11/08/23/mcafee.shows.android.facing.huge.spike.in.malware/). Electronista.com. . Retrieved 2012-01-05.

[168] Dalrymple, Jim (2011-11-16). "Android sees a 472% increase in malware since July" (http://www.loopinsight.com/2011/11/16/android-sees-a-472-increase-in-malware-since-july/). Loopinsight.com. . Retrieved 2012-01-05.
[169] Mobile Malware Development Continues To Rise, Android Leads The Way (http://globalthreatcenter.com/?p=2492).
[170] "The Mother Of All Android Malware Has Arrived" (http://www.androidpolice.com/2011/03/01/the-mother-of-all-android-malware-has-arrived-stolen-apps-released-to-the-market-that-root-your-phone-steal-your-data-and-open-backdoor/). *Android Police*. March 6, 2011. .
[171] Perez, Sarah (2009-02-12). "Android Vulnerability So Dangerous, Owners Warned Not to Use Phone's Web Browser" (http://www.readwriteweb.com/archives/android_vulnerability_so_dangerous_shouldnt_use_web_browser.php). Readwriteweb.com. . Retrieved 2011-08-08.
[172] "Lookout, Retrevo warn of growing Android malware epidemic, note Apple's iOS is far safer" (http://www.appleinsider.com/articles/11/08/03/lookout_retrevio_warn_of_growing_android_malware_epidemic_note_apples_ios_is_far_safer.html). Appleinsider.com. 2011-08-03. . Retrieved 2012-01-05.
[173] "First SMS Trojan detected for smartphones running Android" (http://www.kaspersky.com/news?id=207576158). Kaspersky Lab. . Retrieved 2010-10-18.
[174] "Apple's iOS unaffected by malware as Android exploits surge 76%" (http://www.appleinsider.com/articles/11/08/24/apples_ios_unaffected_by_malware_as_android_exploits_surge_76.html). Appleinsider.com. 2011-08-24. . Retrieved 2012-01-05.
[175] "Security researcher finds code signing flaw that opens door for iOS malware" (http://www.appleinsider.com/articles/11/11/07/newly_found_code_signing_flaw_allows_for_ios_malware.html). Appleinsider.com. 2011-11-07. . Retrieved 2012-01-05.
[176] Apple's iOS more secure than Google's Android, says Symantec (http://www.appleinsider.com/articles/11/06/28/apples_ios_more_secure_than_googles_android_says_symantec.html)
[177] the understatement: Android Orphans: Visualizing a Sad History of Support (http://theunderstatement.com/post/11982112928/android-orphans-visualizing-a-sad-history-of-support)
[178] "Smart phones: how to stay clever in downturn" (http://www.deloitte.co.uk/TMTPredictions/telecommunications/Smartphones-clever-in-downturn.cfm). *Deloitte Telecommunications Predictions*. .
[179] "Android Phones Steal Market Share" (http://bmighty.informationweek.com/mobile/showArticle.jhtml?articleID=224201881). .
[180] "100 Million Club – H1 2010" (http://www.visionmobile.com/blog/2010/10/smart-feature-phones-the-unbalanced-equation-100-million-club-series/). .
[181] "Mobile Phone Development » Blog Archive » Gartner Q3 2010" (http://www.mobilephonedevelopment.com/archives/1149). Mobilephonedevelopment.com. 2010-11-10. . Retrieved 2011-09-07.
[182] Stan Schroeder (2011-02-10). "Gartner: Symbian Is Still the Number One Smartphone Platform" (http://mashable.com/2011/02/10/symbian-number-one-gartner-report/). Mashable.com. . Retrieved 2011-09-07.
[183] "Gartner Says Worldwide Mobile Device Sales to End Users Reached 1.6 Billion Units in 2010; Smartphone Sales Grew 72 Percent in 2010" (http://www.gartner.com/it/page.jsp?id=1543014). Gartner.com. 2011-02-11. . Retrieved 2011-09-07.
[184] "Olswang Predicts Mobile's Impact on TV: The Convergent Revolution" (http://www.tvgenius.net/blog/2011/03/17/mobile-tv-convergence/). Tvgenius.net. 2011-03-17. . Retrieved 2011-09-07.
[185] "Berg: Smartphone shipments grew 74% in 2010" (http://www.bgr.com/2011/03/10/berg-smartphone-shipments-grew-74-in-2010/). *Boy Genius Report*. March 10, 2011. .
[186] Generation App: 62% of Mobile Users 25-34 own Smartphones | Nielsen Wire (http://blog.nielsen.com/nielsenwire/?p=29786).
[187] Market Research | Consumer Market Research - NPD - As Smartphone Prices Fall, Retailers Are Leaving Money on the Table, According to The NPD Group (http://www.npdgroup.com/wps/portal/npd/us/news/pressreleases/pr_111114a)
[188] Apple, With 4 Percent of Handset Market, Captures 52 Percent of Profits (http://www.pcmag.com/article2/0,2817,2395951,00.asp#fbid=SLgcdJun7IU)
[189] "Smartphones killing point-and-shoots, now take almost 1/3 of photos" (http://gigaom.com/2011/12/22/smartphones-killing-point-and-shoots-now-take-almost-13-of-photos/). . Retrieved December 25, 2011.
[190] "Gartner Says Sales of Mobile Devices in Second Quarter of 2011 Grew 16.5 Percent Year-on-Year; Smartphone Sales Grew 74 Percent" (http://www.gartner.com/it/page.jsp?id=1764714). Gartner. 11 August 2011. . Retrieved 20 August 2011.
[191] Tarmo Virki (2011-01-31). "Google topples Symbian from smart phones top spot" (http://www.theglobeandmail.com/news/technology/mobile-technology/google-topples-symbian-from-smartphones-top-spot/article1888557/?cmpid=rss1). Theglobeandmail.com. . Retrieved 2011-09-07.
[192] "Android became clear smartphone leader in first quarter: Canalys" (http://www.reuters.com/article/2011/05/04/us-smartphones-research-idUSTRE7434VV20110504). Reuters.com. 2011-05-04. . Retrieved 2011-09-07.
[193] "Android Takes Over in UK" (http://www.bestsmartphone.com/2011/11/02/android-takes-over-in-uk/). bestsmartphone.com. 2011-11-02. .
[194] http://www.gartner.com/it/page.jsp?id=910112
[195] http://www.gartner.com/it/page.jsp?id=910112
[196] http://www.gartner.com/it/page.jsp?id=1306513
[197] http://www.gartner.com/it/page.jsp?id=1543014
[198] iPass : Mobile Workforce Report (http://mobile-workforce-project.ipass.com/).

[199] iPhone pushes past BlackBerry to top enterprise phone ranks (http://www.appleinsider.com/articles/11/11/16/iphone_pushes_past_blackberry_to_top_enterprise_phone_ranks.html)

[200] In the US Market, iPhone Outperforms Other Mobile Platforms in User Loyalty by a Wide Margin, Android is Second, Blackberry Fourth (http://www.zokem.com/2011/01/in-the-us-market-iphone-outperforms-other-mobile-platforms-in-user-loyalty-by-a-wide-margin-android-is-second-blackberry-fourth/)

[201] iPhone Whips Android, BlackBerry in User Loyalty: Zokem (http://www.eweek.com/c/a/Mobile-and-Wireless/iPhone-Whips-Android-Blackberry-in-User-Loyalty-Zokem-445681/)

[202] Strategy Analytics: Samsung Becomes World's Number One Smartphone Vendor in Q3 2011 - MarketWatch (http://www.marketwatch.com/story/strategy-analytics-samsung-becomes-worlds-number-one-smartphone-vendor-in-q3-2011-2011-10-27)

[203] http://www.gartner.com/it/page.jsp?id=1764714

[204] "Smartphone Sales Will Hit 420 Million In 2011, To Take 28 Percent Of The Total Phone Market" (http://techcrunch.com/2011/07/27/smartphone-sales-will-hit-420-million-in-2011-to-take-28-percent-of-the-total-phone-market/). TechCrunch. 27 July 2011. .

[205] Apple's iPhone accounted for 66% of Q2 smartphone profit among top vendors (http://www.bgr.com/2011/07/29/apples-iphone-accounted-for-66-of-q2-smartphone-profit-among-top-vendors/)

[206] "SA agrees: Apple now top smartphone vendor in the world with 140% growth" (http://www.bgr.com/2011/07/29/sa-agrees-apple-now-top-smartphone-vendor-in-the-world-with-240-growth/). BGR. 29 July 2011. .

[207] "Nokia reports Q2 results: Sells 88.5 million devices and declares net loss of 368 million euros" (http://mobilesyrup.com/2011/07/21/nokia-reports-q2-results-sells-88-5-million-devices-and-declares-net-loss-of-368-million-euros/). Mobilesyrup.com. 2011-07-21. . Retrieved 2011-09-07.

[208] "In U.S. Smartphone Market, Android is Top Operating System, Apple is Top Manufacturer" (http://blog.nielsen.com/nielsenwire/?p=28516). Nielsen. 28 July 2011. . Retrieved 5 September 2011.

[209] "Android Still Dominates Phones, But What About the Rest of Mobile?" (http://www.wired.com/gadgetlab/2011/07/android-ios-platform-share/all/1). Wired. 28 July 2011. . Retrieved 5 September 2011.

[210] iPhone 4 remains top-selling US smartphone despite growing iPhone 5 hype (http://www.appleinsider.com/articles/11/09/06/iphone_4_remains_top_selling_us_smartphone_despite_growing_iphone_5_hype.html)

[211] Previous-gen Apple iPad, iPhone 3GS often outsell new Android devices (http://www.appleinsider.com/articles/11/05/09/previous_gen_apple_ipad_iphone_3gs_often_outsell_new_android_devices.html)

[212] "Samsung Becomes Biggest Smartphone Vendor, as Android's Market Share Grows" (http://www.pcworld.com/article/243861/samsung_becomes_biggest_smartphone_vendor_as_androids_market_share_grows.html). PCWorld. 2011-11-15. . Retrieved 2011-12-10.

[213] Ever-Popular iPhone Named Top Smartphone. Again (http://www.wired.com/gadgetlab/2011/09/iphone-tops-survey-sixth-time/all/1).

[214] Wireless Consumer Smartphone Ratings (Volume 1) (http://www.jdpower.com/Electronics/ratings/wireless-consumer-smartphone-ratings-(volume-1)/)

[215] Wireless Consumer Smartphone Ratings (Volume 2) (http://www.jdpower.com/Electronics/ratings/wireless-consumer-smartphone-ratings-(volume-2)/).

[216] 2011 U.S. Wireless Handset Customer Satisfaction Studies—Vol. 2 (http://www.jdpower.com/news/pressRelease.aspx?ID=2011146)

[217] 2011 Wireless Smartphone and Traditional Mobile Phone Satisfaction Studies-Vol. 1 (http://www.jdpower.com/news/pressrelease.aspx?ID=2011030)

[218] 2010 U.S. Wireless Smartphone and Traditional Mobile Phone Customer Satisfaction Studies-Volume 1 (http://businesscenter.jdpower.com/news/pressrelease.aspx?ID=2010039)

[219] 2009 Wireless Consumer Smartphone and Traditional Mobile Phone Customer Satisfaction Studies (http://businesscenter.jdpower.com/news/pressrelease.aspx?ID=2009082)

[220] Menezes, Gary (2010-09-11). "Symbian OS, Now Fully Open Source" (http://www.watblog.com/2010/02/06/symbian-os-now-fully-open-source). Watblog.com. . Retrieved 2011-09-07.

[221] "Nokia N9 Smart Phone Review" (http://wontek.com/smart-phones/nokia/nokia-n9-smart-phone-review). Wontek.com. 2011-01-01. . Retrieved 2011-09-07.

[222] Lance Whitney, CNET. " Nearly 1 in 5 smartphone owners use check-in services (http://news.cnet.com/8301-1023_3-20062640-93.html?part=rss&subj=news&tag=2547-1_3-0-20)." May 13, 2011. Retrieved May 13, 2011.

[223] Second Screen change the way we watch TV (http://www.good.is/post/double-the-glow-will-second-screen-apps-change-the-way-we-watch-tv/)

[224] explosion of second screen apps (http://www.adweek.com/news/technology/second-screen-apps-explode-132237)

External links

- Freesmartphone.org (http://www.freesmartphone.org) – a collaboration platform for open-source smartphones
- Defining the Smartphone, part 1 (http://www.allaboutsymbian.com/features/item/Defining_the_Smartphone.php), part 2 (http://www.allaboutsymbian.com/features/item/Spy_versus_Spy_No_Smartphone_versus_Smartphone.php) by Steve Litchfield on July 16, 2010, various definitions are treated
- Mozilla Labs: Seabird –Community driven Mobile Phone Concept (http://mozillalabs.com/conceptseries/2010/09/23/seabird/)

Symbian

SYMBIAN	
Company / developer	Accenture on behalf of Nokia[1]
Programmed in	C++[2]
OS family	Embedded operating system
Working state	Active (Receiving updates until at least 2016)[1]
Source model	Proprietary[3]
Initial release	1997 as EPOC32[4]
Latest stable release	Nokia Belle (next release cycle of Symbian^3) / August 24, 2011
Marketing target	Smartphones
Supported platforms	ARM, x86[5]
Kernel type	Microkernel
Default user interface	Avkon[6]
License	Proprietary
Official website	[symbian.nokia.com symbian.nokia.com]

Symbian is a mobile operating system (OS) and computing platform designed for smartphones and currently maintained by Accenture.[7] The Symbian platform is the successor to Symbian OS and Nokia Series 60; unlike Symbian OS, which needed an additional user interface system, Symbian includes a user interface component based on S60 5th Edition. The latest version, Symbian^3, was officially released in Q4 2010, first used in the Nokia N8. In May 2011 an update, Symbian Anna, was officially announced, followed by Nokia Belle (previously Symbian Belle) in August 2011.[8] [9]

Symbian OS was originally developed by Symbian Ltd.[10] It is a descendant of Psion's EPOC and runs exclusively on ARM processors, although an unreleased x86 port existed.

Some estimates indicate that the cumulative number of mobile devices shipped with the Symbian OS up to the end of Q2 2010 is 385 million.[11]

By April 5, 2011, Nokia released Symbian under a new license and converted to a proprietary shared-source model as opposed to an open source project.[3]

On February 11, 2011, Nokia announced that it would migrate away from Symbian to Windows Phone 7. Nokia CEO Stephen Elop announced Nokia's first ever Windows Phones in the Nokia World 2011, the Lumia 800 and Lumia 710. These phones were launched on November 14, 2011.[12] In June 22, 2011 Nokia has made an agreement with Accenture as an outsourcing program. Accenture will provide Symbian based software development and support services to Nokia through 2016 and about 2,800 Nokia employees will be Accenture employees at early October 2011.[13] The transfer was completed on September 30, 2011.[7]

History

The **Symbian** platform was created by merging and integrating software assets contributed by Nokia, NTT DoCoMo, Sony Ericsson and Symbian Ltd., including Symbian OS assets at its core, the S60 platform, and parts of the UIQ and MOAP(S) user interfaces.

In December 2008, Nokia bought Symbian Ltd., the company behind Symbian OS; consequently, Nokia became the major contributor to Symbian's code, since it then possessed the development resources for both the Symbian OS core and the user interface. Since then Nokia has been maintaining its own code repository for the platform development, regularly releasing its development to the public repository.[14] Symbian was intended to be developed by a community led by the Symbian Foundation, which was first announced in June 2008 and which officially launched in April 2009. Its objective was to publish the source code for the entire Symbian platform under the OSI- and FSF-approved Eclipse Public License (EPL). The code was published under EPL on 4 February 2010; Symbian Foundation reported this event to be the largest codebase transitioned to Open Source in history.[15] [16]

However, some important components within Symbian OS were licensed from third parties, which prevented the foundation from publishing the full source under EPL immediately; instead much of the source was published under a more restrictive Symbian Foundation License (SFL) and access to the full source code was limited to member companies only, although membership was open to any organisation.[17]

In November 2010, the Symbian Foundation announced that due to a lack of support from funding members, it would transition to a licensing-only organisation; Nokia announced it would take over the stewardship of the Symbian platform. Symbian Foundation will remain the trademark holder and licensing entity and will only have non-executive directors involved.

On February 11, 2011, Nokia announced a partnership with Microsoft which would see it adopt Windows Phone 7 for smartphones, reducing the number of devices running Symbian over the coming two years.[12] As a consequence, the use of the Symbian platform for building mobile applications dropped rapidly. Research in June 2011 indicated that over 39% of mobile developers using Symbian at the time of publication were planning to abandon the platform.[18]

By April 5, 2011, Nokia ceased to open source any portion of the Symbian software and reduced its collaboration to a small group of pre-selected partners in Japan.[3] Source code released under the EPL remains available in third party repositories.[19] [20]

Version history

Version	Description
EPOC16	EPOC16, originally simply named EPOC, was the operating system developed by Psion in the late 1980s and early 1990s for Psion's "SIBO" (SIxteen Bit Organisers) devices. All EPOC16 devices featured an 8086-family processor and a 16-bit architecture. EPOC16 was a single-user preemptive multitasking operating system, written in Intel 8086 assembler language and C and designed to be delivered in ROM. It supported a simple programming language called Open Programming Language (OPL) and an integrated development environment (IDE) called OVAL. SIBO devices included the: MC200, MC400, Series 3 (1991–98), Series 3a, Series 3c, Series 3mx, Siena, Workabout and Workabout mx. The MC400 and MC200, the first EPOC16 devices, shipped in 1989. EPOC16 featured a primarily 1-bit-per-pixel, keyboard-operated graphical interface[21] — the hardware for which it was designed did not have pointer input. In the late 1990s, the operating system was referred to as **EPOC16** to distinguish it from Psion's then-new EPOC32 OS.

EPOC32 (releases 1 to 5)	The first version of EPOC32, Release 1 appeared on the Psion Series 5 ROM v1.0 in 1997. Later, ROM v1.1 featured Release 3 (Release 2 was never publicly available.) These were followed by the Psion Series 5mx, Revo / Revo plus, Psion Series 7 / netBook and netPad (which all featured Release 5). The EPOC32 operating system, at the time simply referred to as EPOC, was later renamed Symbian OS. Adding to the confusion with names, before the change to Symbian, EPOC16 was often referred to as SIBO to distinguish it from the "new" EPOC. Despite the similarity of the names, EPOC32 and EPOC16 were completely different operating systems, EPOC32 being written in C++ from a new codebase with development beginning during the mid 1990s. EPOC32 was a pre-emptive multitasking, single user operating system with memory protection, which encourages the application developer to separate their program into an engine and an interface. The Psion line of PDAs come with a graphical user interface called EIKON which is specifically tailored for handheld machines with a keyboard (thus looking perhaps more similar to desktop GUIs than palmtop GUIs[22]). However, one of EPOC's characteristics is the ease with which new GUIs can be developed based on a core set of GUI classes, a feature which has been widely explored from Ericsson R380 and onwards. EPOC32 was originally developed for the ARM family of processors, including the ARM7, ARM9, StrongARM and Intel's XScale, but can be compiled towards target devices using several other processor types. During the development of EPOC32, Psion planned to license EPOC to third-party device manufacturers, and spin off its software division as Psion Software. One of the first licensees was the short-lived *Geofox*, which halted production with less than 1,000 units sold. Ericsson marketed a rebranded Psion Series 5mx called the *MC218*, and later created the EPOC Release 5.1 based smartphone, the *R380*. Oregon Scientific also released a budget EPOC device, the *Osaris* (notable as the only EPOC device to ship with Release 4). Work started on the 32-bit version in late 1994. The Series 5 device, released in June 1997, used the first iterations of the EPOC32 OS, codenamed "Protea", and the "Eikon" graphical user interface. The Oregon Scientific Osaris was the only PDA to use the ER4. The Psion Series 5mx, Psion Series 7, Psion Revo, Diamond Mako, Psion netBook and Ericsson MC218 were released in 1999 using ER5. A phone project was announced at CeBIT, the Phillips Illium/Accent, but did not achieve a commercial release. This release has been retrospectively dubbed Symbian OS 5. The first phone using ER5u, the Ericsson R380 was released in November 2000. It was not an 'open' phone – software could not be installed. Notably, a number of never-released Psion prototypes for next generation PDAs, including a Bluetooth Revo successor codenamed "Conan" were using ER5u. The 'u' in the name refers to the fact that it supported Unicode. In June 1998, Psion Software became Symbian Ltd., a major joint venture between Psion and phone manufacturers Ericsson, Motorola, and Nokia. As of Release 6, EPOC became known simply as Symbian OS.
Symbian OS 6.0 and 6.1	The OS was renamed Symbian OS and was envisioned as the base for a new range of smartphones. This release is sometimes called ER6. Psion gave 130 key staff to the new company and retained a 31% shareholding in the spin-off. The first 'open' Symbian OS phone, the Nokia 9210 Communicator, was released in June 2001. Bluetooth support was added. Almost 500,000 Symbian phones were shipped in 2001, rising to 2.1 million the following year. Development of different UIs was made generic with a "reference design strategy" for either 'smartphone' or 'communicator' devices, subdivided further into keyboard- or tablet-based designs. Two reference UIs (DFRDs or Device Family Reference Designs) were shipped – Quartz and Crystal. The former was merged with Ericsson's 'Ronneby' design and became the basis for the UIQ interface; the latter reached the market as the Nokia Series 80 UI. Later DFRDs were Sapphire, Ruby, and Emerald. Only Sapphire came to market, evolving into the Pearl DFRD and finally the Nokia Series 60 UI, a keypad-based 'square' UI for the first true smartphones. The first one of them was the Nokia 7650 smartphone (featuring Symbian OS 6.1), which was also the first with a built-in camera, with VGA (0.3 Mpx = 640×480) resolution. Despite these efforts to be generic, the UI was clearly split between competing companies: Crystal or Sapphire was Nokia, Quartz was Ericsson. DFRD was abandoned by Symbian in late 2002, as part of an active retreat from UI development in favour of 'headless' delivery. Pearl was given to Nokia, Quartz development was spun off as UIQ Technology AB, and work with Japanese firms was quickly folded into the MOAP standard.

Symbian OS 7.0 and 7.0s	First shipped in 2003. This is an important Symbian release which appeared with all contemporary user interfaces including UIQ (Sony Ericsson P800, P900, P910, Motorola A925, A1000), Series 80 (Nokia 9300, 9500), Series 90 (Nokia 7710), Series 60 (Nokia 3230, 6260, 6600, 6670, 7610) as well as several FOMA phones in Japan and Siemens SX1(VGA Camera, MMC card, Bluetooth, Infraport, radio) - the first and the last symbian phone from Siemens. It also added EDGE support and IPv6. Java support was changed from pJava and JavaPhone to one based on the Java ME standard. One million Symbian phones were shipped in Q1 2003, with the rate increasing to one million a month by the end of 2003. Symbian OS 7.0s was a version of 7.0 special adapted to have greater backward compatibility with Symbian OS 6.x, partly for compatibility between the Communicator 9500 and its predecessor the Communicator 9210. In 2004, Psion sold its stake in Symbian. The same year, the first worm for mobile phones using Symbian OS, *Cabir*, was developed, which used Bluetooth to spread itself to nearby phones. See Cabir and Symbian OS threats.
Symbian OS 8.0	First shipped in 2004, one of its advantages would have been a choice of two different kernels (EKA1 or EKA2). However, the EKA2 kernel version did not ship until Symbian OS 8.1b. The kernels behave more or less identically from user-side, but are internally very different. EKA1 was chosen by some manufacturers to maintain compatibility with old device drivers, while EKA2 was a real-time kernel. 8.0b was deproductised in 2003. Also included were new APIs to support CDMA, 3G, two-way data streaming, DVB-H, and OpenGL ES with vector graphics and direct screen access.
Symbian OS 8.1	An improved version of 8.0, this was available in 8.1a and 8.1b versions, with EKA1 and EKA2 kernels respectively. The 8.1b version, with EKA2's single-chip phone support but no additional security layer, was popular among Japanese phone companies desiring the real-time support but not allowing open application installation. The first and maybe the most famous smartphone featuring Symbian OS 8.1a was Nokia N90 in 2005, Nokia's first in Nseries.
Symbian OS 9.0	Symbian OS 9.0 was used for internal Symbian purposes only. It was de-productised in 2004. 9.0 marked the end of the road for EKA1. 8.1a is the final EKA1 version of Symbian OS. Symbian OS has generally maintained reasonable binary code compatibility. In theory the OS was BC from ER1-ER5, then from 6.0 to 8.1b. Substantial changes were needed for 9.0, related to tools and security, but this should be a one-off event. The move from requiring ARMv4 to requiring ARMv5 did not break backwards compatibility.
Symbian OS 9.1	Released early 2005. It includes many new security related features, including platform security module facilitating mandatory code signing. The new ARM EABI binary model means developers need to retool and the security changes mean they may have to recode. S60 platform 3rd Edition phones have Symbian OS 9.1. Sony Ericsson is shipping the M600 and P990 based on Symbian OS 9.1. The earlier versions had a defect where the phone hangs temporarily after the owner sent a large number of SMS'es. However, on 13 September 2006, Nokia released a small program to fix this defect.[23] Support for Bluetooth 2.0 was also added. Symbian 9.1 introduced capabilities and a Platform Security framework. To access certain APIs, developers have to sign their application with a digital signature. Basic capabilities are user-grantable and developers can self-sign them, while more advanced capabilities require certification and signing via the Symbian Signed [24] program, which uses independent 'test houses' and phone manufacturers for approval. For example, file writing is a user-grantable capability while access to Multimedia Device Drivers require phone manufacturer approval. A TC TrustCenter ACS Publisher ID certificate is required by the developer for signing applications.
Symbian OS 9.2	Released Q1 2006. Support for OMA Device Management 1.2 (was 1.1.2). Vietnamese language support. S60 3rd Edition Feature Pack 1 phones have Symbian OS 9.2. Nokia phones with Symbian OS 9.2 OS include the Nokia E71, Nokia E90, Nokia N95, Nokia N82, Nokia N81 and Nokia 5700.
Symbian OS 9.3	Released on 12 July 2006. Upgrades include improved memory management and native support for Wifi 802.11, HSDPA. The Nokia E72, Nokia 5730 XpressMusic, Nokia N79, Nokia N96, Nokia E52, Nokia E75, Nokia 5320 XpressMusic, Sony Ericsson P1 and others feature Symbian OS 9.3.
Symbian OS 9.4	Announced in March 2007. Provides the concept of demand paging which is available from v9.3 onwards. Applications should launch up to 75% faster. Additionally, SQL support is provided by SQLite. Ships with the Samsung i8910 Omnia HD, Nokia N97, Nokia N97 mini, Nokia 5800 XpressMusic, Nokia 5530 XpressMusic, Nokia 5228, Nokia 5230, Nokia 5233, Nokia 5235, Nokia C6-00, Nokia X6, Sony Ericsson Satio, Sony Ericsson Vivaz and Sony Ericsson Vivaz Pro,Micromax x265. Used as the basis for Symbian^1, the first Symbian platform release. The release is also better known as S60 5th edition, as it is the bundled interface for the OS.

Symbian^3 (Symbian OS 9.5) and Symbian Anna	Symbian^3 is a big improvement over previous S60 5th Edition and features single touch menus in the user interface, as well as new Symbian OS kernel with hardware-accelerated graphics; further improvements will come in the first half of 2011 including portrait qwerty keyboard, a new browser and split-screen text input. Nokia announced that updates to Symbian^3 interface will be delivered gradually, as they are available; Symbian^4, the previously planned major release, is now discontinued and some of its intended features will be incorporated into Symbian^3 in successive releases, starting with Symbian Anna.
Nokia Belle	In the summer of 2011 videos showing an early leaked version of Nokia Belle (formally Symbian Belle) running on a Nokia N8 were published on YouTube. On August 24, 2011, Nokia announced Nokia Belle officially for three new smartphones, the Nokia 603, Nokia 700, and Nokia 701. They also announced that Belle would be coming to all existing Symbian^3 devices in February 2012.[25] Nokia officially renamed Symbian Belle to Nokia Belle in a company blog post.[26] Nokia Belle adds to the Anna improvements with a pull-down status/notification bar, deeper near field communication integration, free-form re-sizable homescreen widgets, and six homescreens instead of the previous three.

Features

User interface

Symbian has had a native graphics toolkit since its inception, known as AVKON (formerly known as Series 60). S60 was designed to be manipulated by a keyboard-like interface metaphor, such as the ~15-key augmented telephone keypad, or the mini-QWERTY keyboards. AVKON-based software is binary-compatible with Symbian versions up to and including Symbian^3.

Symbian^3 includes the Qt framework, which is now the recommended user interface toolkit for new applications. Qt can also be installed on older Symbian devices.

Symbian^4 was planned to introduce a new GUI library framework specifically designed for a touch-based interface, known as "UI Extensions for Mobile" or UIEMO (internal project name "Orbit"), which was built on top of Qt Widget; a preview was released in January 2010, however in October 2010 Nokia announced that Orbit/UIEMO has been cancelled.

Nokia currently recommends that developers use Qt Quick with QML, the new high-level declarative UI and scripting framework for creating visually rich touchscreen interfaces that allows development for both Symbian and MeeGo; it will be delivered to existing Symbian^3 devices as a Qt update. When more applications gradually feature a user interface reworked in Qt, the legacy S60 framework (AVKON) will be deprecated and no longer included with new devices at some point, thus breaking binary compatibility with older S60 applications.[27] [28]

Browser

Symbian^3 and earlier have a native WebKit based browser; indeed, Symbian was the first mobile platform to make use of WebKit (in June 2005).[29] Some older Symbian models have Opera Mobile as their default browser.

Nokia released a new browser with the release of Symbian Anna with improved speed and an improved user interface.[30]

Multiple language support

Symbian has strong localization support enabling manufacturers and 3rd party application developers to localize their Symbian based products in order to support global distribution.

Current Symbian release (Symbian Belle) has support for 48 languages, which Nokia makes available on device in language packs (set of languages which cover the languages commonly spoken in the area where the device variant is intended to be sold). All language packs have in common English (or a locally relevant dialect of it).

The supported languages [with dialects] (and scripts) in Symbian Belle are:

- Arabic (Arabic),
- Basque (Latin),
- Bulgarian (Cyrillic),
- Catalan (Latin),
- Chinese [PRC] (Simplified Chinese),
- Chinese [Hong Kong] (Traditional Chinese),
- Chinese [Taiwan] (Traditional Chinese),
- Croatian (Latin),
- Czech (Latin),
- Danish (Latin),
- Dutch (Latin),
- English [UK] (Latin),
- English [US] (Latin),
- Estonian (Latin),
- Finnish (Latin),
- French (Latin),
- French [Canadian] (Latin),
- Galician (Latin),
- German (Latin),
- Greek (Greek),
- Hebrew (Hebrew),
- Hungarian (Latin),
- Icelandic (Latin),
- Indonesian [Bahasa Indonesian] (Latin),
- Italian (Latin),
- Kazakh (Cyrillic),
- Latvian (Latin),
- Lithuanian (Latin),
- Malay [Bahasa Malaysia] (Latin),
- Norwegian (Latin),
- Persian [Farsi] (Arabic),
- Polish (Latin),
- Portuguese (Latin),
- Portuguese [Brazilian] (Latin),
- Romanian (Latin),
- Russian (Cyrillic),
- Serbian (Latin),
- Slovak (Latin),
- Slovene (Latin),
- Spanish (Latin),
- Spanish [Latin America] (Latin),
- Swedish (Latin),
- Tagalog [Filipino] (Latin),
- Thai (Thai),
- Turkish (Latin),
- Ukrainian (Cyrillic),
- Urdu (Arabic),
- Vietnamese (Latin).

Symbian Belle marks the introduction of Kazakh, while Korean is no longer supported.

Application development

From 2010, Symbian switched to using standard C++ with Qt as the main SDK, which can be used with either Qt Creator or Carbide.c++. Qt supports the older Symbian/S60 3rd (starting with Feature Pack 1, aka S60 3.1) and Symbian/S60 5th Edition (aka S60 5.0) releases, as well as the new Symbian platform. It also supports Maemo and MeeGo, Windows, Linux and Mac OS X.[31] [32]

Alternative application development can be done using Python (see Python for S60), Adobe Flash Lite or Java ME.

Symbian OS previously used a Symbian specific C++ version, along with Carbide.c++ integrated development environment (IDE), as the native application development environment.

Web Runtime (WRT) is a portable application framework that allows creating widgets on the S60 Platform; it is an extension to the S60 WebKit based browser that allows launching multiple browser instances as separate JavaScript applications.[33] [34]

Application development

Qt

As of 2010, the SDK for Symbian is standard C++, using Qt. It can be used with either Qt Creator, or Carbide (the older IDE previously used for Symbian development).[31] [35] A phone simulator allows testing of Qt apps. Apps compiled for the simulator are compiled to native code for the development platform, rather than having to be emulated.[36] Application development can either use C++ or QML.

Symbian C++

It is also possible to develop using Symbian C++, although it is not a standard implementation. Before the release of the Qt SDK, this was the standard development environment. There were multiple platforms based on Symbian OS that provided software development kit (SDKs) for application developers wishing to target Symbian OS devices, the main ones being UIQ and S60. Individual phone products, or families, often had SDKs or SDK extensions downloadable from the maker's website too.

The SDKs contain documentation, the header files and library files needed to build Symbian OS software, and a Windows-based emulator ("WINS"). Up until Symbian OS version 8, the SDKs also included a version of the GNU Compiler Collection (GCC) compiler (a cross-compiler) needed to build software to work on the device.

Symbian OS 9 and the Symbian platform use a new application binary interface (ABI) and needed a different compiler. A choice of compilers is available including a newer version of GCC (see external links below).

Unfortunately, Symbian C++ programming has a steep learning curve, as Symbian C++ requires the use of special techniques such as descriptors, active objects and the cleanup stack. This can make even relatively simple programs harder to implement than in other environments. Moreover, it was questionable whether these techniques, such as the memory management paradigm, were actually beneficial. It is possible that the techniques, developed for the much more restricted mobile hardware of the 1990s, simply caused unnecessary complexity in source code because programmers are required to concentrate on low-level routines instead of more application-specific features. As of 2010, these issues are no longer the case when using standard C++, with the Qt SDK.

Symbian C++ programming is commonly done with an integrated development environment (IDE). For earlier versions of Symbian OS, the commercial IDE CodeWarrior for Symbian OS was favoured. The CodeWarrior tools were replaced during 2006 by Carbide.c++, an Eclipse-based IDE developed by Nokia. Carbide.c++ is offered in four different versions: Express, Developer, Professional, and OEM, with increasing levels of capability. Fully featured software can be created and released with the Express edition, which is free. Features such as UI design, crash debugging etc. are available in the other, charged-for, editions. Microsoft Visual Studio 2003 and 2005 are also supported via the Carbide.vs plugin.

Other languages

Symbian devices can also be programmed using Python, Java ME, Flash Lite, Ruby, .NET, Web Runtime (WRT) Widgets and Standard C/C++.[37]

Visual Basic programmers can use NS Basic to develop apps for S60 3rd Edition and UIQ 3 devices.

In the past, Visual Basic, Visual Basic .NET, and C# development for Symbian were possible through AppForge Crossfire, a plugin for Microsoft Visual Studio. On 13 March 2007 AppForge ceased operations; Oracle purchased the intellectual property, but announced [38] that they did not plan to sell or provide support for former AppForge products. Net60 [39], a .NET compact framework for Symbian, which is developed by redFIVElabs, is sold as a commercial product. With Net60, VB.NET and C# (and other) source code is compiled into an intermediate language (IL) which is executed within the Symbian OS using a just-in-time compiler. (As of 18/1/10 RedFiveLabs has ceased development of Net60 with this announcement on their landing page: "At this stage we are pursuing some options to sell the IP so that Net60 may continue to have a future".)

There is also a version of a Borland IDE for Symbian OS. Symbian OS development is also possible on Linux and Mac OS X using tools and methods developed by the community, partly enabled by Symbian releasing the source code for key tools. A plugin that allows development of Symbian OS applications in Apple's Xcode IDE for Mac OS X was available.[40]

Java ME applications for Symbian OS are developed using standard techniques and tools such as the Sun Java Wireless Toolkit (formerly the J2ME Wireless Toolkit). They are packaged as JAR (and possibly JAD) files. Both CLDC and CDC applications can be created with NetBeans. Other tools include SuperWaba, which can be used to build Symbian 7.0 and 7.0s programs using Java.

Nokia S60 phones can also run Python scripts when the interpreter Python for S60 is installed, with a custom made API that allows for Bluetooth support and such. There is also an interactive console to allow the user to write Python scripts directly from the phone.

Deployment

Once developed, Symbian applications need to find a route to customers' mobile phones. They are packaged in SIS files which may be installed over-the-air, via PC connect, Bluetooth or on a memory card. An alternative is to partner with a phone manufacturer and have the software included on the phone itself. Applications must be Symbian Signed [24] for Symbian OS 9.x in order to make use of certain capabilities (system capabilities, restricted capabilities and device manufacturer capabilities).[41] Applications can now be signed for free.[42]

Architecture

Technology domains and packages

Symbian's design is subdivided into **technology domains**,[43] each of which comprises a number of software **packages**.[44] Each technology domain has its own roadmap, and the Symbian Foundation has a team of technology managers who manage these technology domain roadmaps.

Every package is allocated to exactly one technology domain, based on the general functional area to which the package contributes and by which it may be influenced. By grouping related packages by themes, the Symbian Foundation hopes to encourage a strong community to form around them and to generate discussion and review.

The Symbian System Model[45] illustrates the scope of each of the technology domains across the platform packages.

Packages are owned and maintained by a package owner, a named individual from an organization member of the Symbian Foundation, who accepts code contributions from the wider Symbian community and is responsible for package.

Symbian kernel

The Symbian kernel (EKA2) supports sufficiently fast real-time response to build a single-core phone around it—that is, a phone in which a single processor core executes both the user applications and the signalling stack.[46] The real-time kernel has a microkernel architecture containing only the minimum, most basic primitives and functionality, for maximum robustness, availability and responsiveness. It has been termed a nanokernel, because it needs an extended kernel to implement any other abstractions. It contains a scheduler, memory management and device drivers, with networking, telephony and file system support services in the OS Services Layer or the Base Services Layer. The inclusion of device drivers means the kernel is not a *true* microkernel.

Design

Symbian features pre-emptive multitasking and memory protection, like other operating systems (especially those created for use on desktop computers). EPOC's approach to multitasking was inspired by VMS and is based on asynchronous server-based events.

Symbian OS was created with three systems design principles in mind:

1. the integrity and security of user data is paramount
2. user time must not be wasted
3. all resources are scarce

To best follow these principles, Symbian uses a microkernel, has a request-and-callback approach to services, and maintains separation between user interface and engine. The OS is optimised for low-power battery-based devices and for ROM-based systems (e.g. features like XIP and re-entrancy in shared libraries). Applications, and the OS itself, follow an object-oriented design: Model-view-controller (MVC).

Later OS iterations diluted this approach in response to market demands, notably with the introduction of a real-time kernel and a platform security model in versions 8 and 9.

There is a strong emphasis on conserving resources which is exemplified by Symbian-specific programming idioms like descriptors and a cleanup stack. Similar methods exist to conserve disk space, though disks on Symbian devices are usually flash memory. Further, all Symbian programming is event-based, and the central processing unit (CPU) is switched into a low power mode when applications are not directly dealing with an event. This is done via a programming idiom called active objects. Similarly the Symbian approach to threads and processes is driven by reducing overheads.

Operating system

The All over Model contains the following layers, from top to bottom:

- UI Framework Layer
- Application Services Layer
  - Java ME
- OS Services Layer
  - generic OS services
  - communications services
  - multimedia and graphics services
  - connectivity services
- Base Services Layer
- Kernel Services & Hardware Interface Layer

The Base Services Layer is the lowest level reachable by user-side operations; it includes the File Server and User Library, a Plug-In Framework which manages all plug-ins, Store, Central Repository, DBMS and cryptographic services. It also includes the Text Window Server and the Text Shell: the two basic services from which a completely functional port can be created without the need for any higher layer services.

Symbian has a microkernel architecture, which means that the minimum necessary is within the kernel to maximise robustness, availability and responsiveness. It contains a scheduler, memory management and device drivers, but other services like networking, telephony and filesystem support are placed in the OS Services Layer or the Base Services Layer. The inclusion of device drivers means the kernel is not a *true* microkernel. The EKA2 real-time kernel, which has been termed a nanokernel, contains only the most basic primitives and requires an extended kernel to implement any other abstractions.

Symbian is designed to emphasise compatibility with other devices, especially removable media file systems. Early development of EPOC led to adopting FAT as the internal file system, and this remains, but an object-oriented persistence model was placed over the underlying FAT to provide a POSIX-style interface and a streaming model. The internal data formats rely on using the same APIs that create the data to run all file manipulations. This has resulted in data-dependence and associated difficulties with changes and data migration.

There is a large networking and communication subsystem, which has three main servers called: ETEL (EPOC telephony), ESOCK (EPOC sockets) and C32 (responsible for serial communication). Each of these has a plug-in scheme. For example, ESOCK allows different ".PRT" protocol modules to implement various networking protocol schemes. The subsystem also contains code that supports short-range communication links, such as Bluetooth, IrDA and USB.

There is also a large volume of user interface (UI) Code. Only the base classes and substructure were contained in Symbian OS, while most of the actual user interfaces were maintained by third parties. This is no longer the case. The three major UIs — S60, UIQ and MOAP — were contributed to Symbian in 2009. Symbian also contains graphics, text layout and font rendering libraries.

All native Symbian C++ applications are built up from three framework classes defined by the application architecture: an application class, a document class and an application user interface class. These classes create the fundamental application behaviour. The remaining needed functions, the application view, data model and data interface, are created independently and interact solely through their APIs with the other classes.

Many other things do not yet fit into this model — for example, SyncML, Java ME providing another set of APIs on top of most of the OS and multimedia. Many of these are frameworks, and vendors are expected to supply plug-ins to these frameworks from third parties (for example, Helix Player for multimedia codecs). This has the advantage that the APIs to such areas of functionality are the same on many phone models, and that vendors get a lot of flexibility. But it means that phone vendors needed to do a great deal of integration work to make a Symbian OS phone.

Symbian includes a reference user-interface called "TechView." It provides a basis for starting customisation and is the environment in which much Symbian test and example code runs. It is very similar to the user interface from the Psion Series 5 personal organiser and is not used for any production phone user interface.

Devices and feature comparison

On 16 November 2006, the 100 millionth smartphone running the OS was shipped.[47] As of 21 July 2009, more than 250 million devices running Symbian OS had been shipped.[48]

- The Nokia S60 interface is used in various phones, the first being the Nokia 7650. The Nokia N-Gage and Nokia N-Gage QD gaming/smartphone combos are also S60 platform devices. It was also used on other manufacturers' phones such as the Siemens SX1 and Samsung SGH-Z600. Recently, more advanced devices using S60 include the Nokia 6xxx, the Nseries (except Nokia N8xx and N9xx), the Eseries and some models of the Nokia XpressMusic mobiles.
- Fujitsu, Mitsubishi, Sony Ericsson and Sharp developed phones for NTT DoCoMo in Japan, using an interface developed specifically for DoCoMo's FOMA "Freedom of Mobile Access" network brand. This UI platform is called MOAP "Mobile Oriented Applications Platform" and is based on the UI from earlier Fujitsu FOMA models. The user cannot install new C++ applications.

User interfaces that run on or are based on Symbian OS include:

- S60, formerly Series 60, used by Nokia and others
- Series 80, previously used by Nokia
- Series 90, previously used by Nokia
- UIQ, previously used by Sony-Ericsson
- MOAP, Mobile Oriented Applications Platform, used by NTT DoCoMo's FOMA service
- OPP, successor of MOAP, used on NTT DoCoMo's FOMA phone

Versions that are actively marketed as of September 2011 are Symbian^3 (and its updated Symbian Anna and Nokia Belle variants), Symbian^2, Symbian^1 (previously known as Series 60 5th Edition), and Series 60 3rd Edition Feature Pack 2. For features of older versions, see history of Symbian. Note that the operating system supporting a certain feature does not imply that all devices running on it have that feature available, especially if it involves expensive hardware, such as HDMI output.

Feature	Symbian^3/Anna/Belle	Symbian^2[49]	Symbian^1/Series 60 5th Edition	Series 60 3rd Edition Feature Pack 2	Series 80
Year released	2010 (Symbian^3), 2011 (Symbian Anna, Nokia Belle)	2010	2008	2008	
Company	Symbian Foundation	Symbian Foundation	Symbian Foundation	Symbian Foundation	Symbian Foundation
Symbian OS version	9.5 (Symbian^3)		9.4	9.3	
Series 60 version	5.2 (Symbian^3)[50]	5.1	5th Edition	3rd Edition Feature Pack 2	N/A
Touch input support	Yes	Yes	Yes	No	No
Multi touch input support	Yes		No	No	No
Number of customizable home screens	Three to six (Five on Nokia E6, six on Nokia Belle)		One	Two	One
Wi-Fi version support	B, G, N		B, G	B, G	B, G
USB on the go support	Yes		No	No	
DVB-H support	Yes, with extra headset[51]		Yes, with extra headset	Yes, with extra headset	
Short range FM transmitter support	Yes		Yes	Yes	No
FM radio support	Yes		Yes	Yes	No
Feature	**Symbian^3/Anna/Belle**	**Symbian^2**	**Symbian^1/Series 60 5th Edition**	**Series 60 3rd Edition Feature Pack 2**	**Series 80**
Adobe Flash support	Yes, Flash Lite native version 4.0, upgradable		Yes, Flash Lite native version 3.1, upgradable	Yes, Flash Lite native version 3.1, upgradable	No
Microsoft Silverlight support	Yes[52]		Yes[52]	Yes[52]	No
OpenGL ES support	Yes, version 2.0				No
SQLite support	Yes		Yes	Yes[53]	
CPU architecture support	ARM	SH-Mobile	ARM	ARM	ARM
Programmed in	C++, Qt		C++, Qt	C++, Qt	

License	Eclipse Public License; Since March 31, 2011: Nokia Symbian License 1.0	proprietary SFL license, while some portions of source code are EPL licensed.			
Public issues list	No more				
Package manager	.sis, .sisx		.sis, .sisx	.sis, .sisx	.sis, .sisx
Non English languages support	Yes	Yes, Japanese	Yes	Yes	Yes
Underlining spell checker	Yes		Yes	Yes	
Keeps state on shutdown or crash	No		No	No	No
Internal search	Yes		Yes	Yes	Yes
Proxy server	Yes		Yes	Yes	Yes
On-device encryption	Yes		Yes	Yes	
Cut, copy, and paste support	Yes		Yes	Yes	Yes
Undo	No		No	No	Yes
Default Web Browser for S60, WebKit engine	version 7.2, engine version 525 (Symbian^3);[54] version 7.3, engine version 533.4 (Symbian Anna)		version 7.1.4, engine version 525; version 7.3, engine version 533.4 (for 9 selected units after firmware updates released in summer 2011)	engine version 413 (Nokia N79)	N/A
third-party software store	Ovi store		Ovi store	Ovi store	
Email sync protocol support	POP3, IMAP		POP3, IMAP	POP3, IMAP	POP3, IMAP
Feature	**Symbian^3/Anna/Belle**	**Symbian^2**	**Symbian^1/Series 60 5th Edition**	**Series 60 3rd Edition Feature Pack 2**	**Series 80**
Push alerts	Yes		Yes	Yes	Yes
Voice recognition	Yes	Yes	Yes	Yes	
Tethering	USB, Bluetooth; mobile Wi-Fi hotspot, with third-party software		USB, Bluetooth; mobile Wi-Fi hotspot, with third-party software	USB, Bluetooth; mobile Wi-Fi hotspot, with third-party software	USB, Bluetooth;
Text, document support	Mobile Office Applications, PDF	Mobile Office Applications, PDF	Mobile Office Applications, PDF	Mobile Office Applications, PDF	Mobile Office Applications, PDF

Audio playback	All	wma, aac	All	All	wav, mp3
Video playback	H.263, H.264, WMV, MPEG4, MPEG4@ HD 720p 25–30 frame/s, MKV, DivX, XviD		H.263, WMV, MPEG4, 3GPP, 3GPP2	H.263, WMV, MPEG4, 3GPP, 3GPP2	H.263, 3GPP, 3GPP2
Turn-by-turn GPS	third-party software, free global Nokia Ovi Maps, works offline		third-party software, free global Nokia Ovi Maps, works offline	third-party software, free global Nokia Ovi Maps, works offline	third-party software
Video out	Nokia AV, PAL, NTSC, HDMI	HDMI, and	Nokia AV, PAL, NTSC	Nokia AV, PAL, NTSC	No
Multitasking	Yes		Yes	Yes	Yes
Desktop interactive widgets	Yes	Yes	Yes	No	
Integrated hardware keyboard	Yes	Yes	Yes	Yes	Yes
Bluetooth keyboard	Yes		Yes	Yes	Yes
Video conference front video camera	Yes	Yes	Yes	Yes	Yes
Can share data via Bluetooth with all devices	Yes	Yes	Yes	Yes	Yes
Skype, third-party software	Yes[55]		Yes[55]	Yes[55]	
Facebook IM chat	Yes		Yes	Yes	
Secure Shell (SSH)	Yes, third-party software		Yes, third-party software	Yes, third-party software	
OpenVPN	No, Nokia VPN can be used		No, Nokia VPN can be used	No, Nokia VPN can be used	Yes, third-party software
Remote frame buffer	?				
Screenshot	Yes, third-party software[56]		Yes, third-party software[56]	Yes, third-party software[56]	Yes
GPU acceleration	Yes				No

Official SDK platform(s)	Cross-platform, Windows (preferred is Qt), Carbide.c++, Java ME, Web Runtime Widgets (WRT), Flash lite, Python for Symbian		Cross-platform, Windows (preferred is Qt), Carbide.c++, Java ME, Web Runtime Widgets (WRT), Flash lite, Python for Symbian	Cross-platform, Windows (preferred is Qt), Carbide.c++, Java ME, Web Runtime Widget (WRT), Flash lite, Python for Symbian	Cross-platform, Windows (preferred is Qt), Carbide.c++, Java ME, third-party software (OPL)
Feature	**Symbian^3/Anna/Belle**	**Symbian^2**	**Symbian^1/Series 60 5th Edition**	**Series 60 3rd Edition Feature Pack 2**	**Series 80**
First device(s)	Nokia N8 (Symbian^3), Nokia C7 (Symbian^3), Nokia X7, Nokia E6 (Anna), Nokia 603, Nokia 700, Nokia 701 (Belle)	NTT DOCOMO STYLE Series F-07B	Nokia 5800 (October 2, 2008)	Nokia N96, Nokia N78, Nokia 6210 Navigator and Nokia 6220 Classic (February 11, 2008)	Nokia 9210
Devices	Nokia N8, Nokia C6-01, Nokia C7-00, Nokia E7-00, Nokia E6, Nokia X7, Nokia 500, Nokia 603, Nokia 700, Nokia 701	NTT DoCoMo: F-06B*,[57] F-07B*,[57] F-08B*,[57] SH-07B†,[57] F-10B,[58] Raku-Raku Phone 7,[58] F-01C*,[59] F-02C*,[59] F-03C*,[59] F-04C*,[59] F-05C*,[59] SH-01C†,[59] SH-02C†,[59] SH-04C†,[59] SH-05C†,[59] SH-06C†,[59] Touch Wood SH-08C†[59]	Nokia 5228, Nokia 5230, Nokia 5233, Nokia 5235, Nokia 5250, Nokia 5530 XpressMusic, Nokia 5800 XpressMusic, Nokia 5800 Navigation Edition, Nokia C5-03, Nokia C6-00, Nokia N97, Nokia N97 mini, Nokia X6, Samsung i8910 Omnia HD,[60] Sony Ericsson Satio, Sony Ericsson Vivaz, Sony Ericsson Vivaz Pro	Nokia 5320 XpressMusic, Nokia 5630 XpressMusic, Nokia 5730 XpressMusic, Nokia 6210 Navigator, Nokia 6220 Classic, Nokia 6650 fold, Nokia 6710 Navigator, Nokia 6720 Classic, Nokia 6730 Classic, Nokia 6760 Slide, Nokia 6790 Surge, Nokia E5-00, Nokia E52, Nokia E55, Nokia E71, Nokia E72, Nokia E75, Nokia N78, Nokia N79, Nokia N85, Nokia N86 8MP, Nokia N96, Nokia X5, Samsung GT-i8510 (INNOV8), Samsung GT-I7110, Samsung SGH-L870, Nokia C5-00	Nokia 9210, Nokia 9300, Nokia 9300i, Nokia 9500
Feature	**Symbian^3/Anna/Belle**	**Symbian^2**	**Symbian^1/Series 60 5th Edition**	**Series 60 3rd Edition Feature Pack 2**	**Series 80**

* manufactured by Fujitsu

† manufactured by Sharp

Market share and competition

In the number of "smart mobile device" sales, Symbian devices were the market leaders for 2010. Statistics showed that Symbian devices formed a 37.6% share of smart mobile devices sold, with Android having 22.7%, RIM having 16%, and Apple having 15.7% (via iOS).[61]

Prior reports on device shipments as published in February 2010 showed that the Symbian devices formed a 47.2% share of the smart mobile devices shipped in 2009, with RIM having 20.8%, Apple having 15.1% (via iOS), Microsoft having 8.8% (via Windows CE and Windows Mobile) and Android having 4.7%.[62] Other competitors include webOS, Qualcomm's BREW, SavaJe, Linux and MontaVista Software.

Symbian has lost market share over the years as the market has dramatically grown, with new competing platforms entering the market, though it's sales have increased during the same timeframe. E.g., although Symbian's share of the global smartphone market dropped from 52.4% in 2008 to 47.2% in 2009, shipments of Symbian devices grew 4.8%, from 74.9 million units to 78.5 million units.[62] From Q2 2009 to Q2 2010, shipments of Symbian devices grew 41.5%, by 8.0 million units, from 19,178,910 units to 27,129,340; compared to an increase of 9.6 million units for Android, 3.3 million units for RIM, and 3.2 million units for Apple.[63] In 2006, Symbian had 73% of the smartphone market,[64] compared with 22.1% of the market in the second quarter of 2011.[65] Over the course of 2009–2011, Nokia, Motorola, Samsung, LG, and Sony Ericsson announced their withdrawal from Symbian in favour of alternative platforms including Google's Android, Microsoft's Windows Phone, and Samsung's bada.[66] [67] [68] [69]

Criticisms

The users of Symbian in the countries with non-Latin alphabets (such as Russia, Ukraine and others) have been criticizing the complicated method of language switching for many years.[70] For example, if a user wants to type a Latin letter, he must call the menu, click the languages item, choose the English language between many other languages by arrow keys and then press the 'OK' button. After typing the Latin letter, the user must repeat that procedure to return to his native keyboard. This method slows down the typing significantly. In touch-phones and QWERTY phones the procedure is slightly different but remains time-consuming. All other mobile operating systems as well as Nokia's S40 phones enable switching between two initially selected languages by one click or by one gesture.

Early versions of the firmware for the original Nokia N97, running on Symbian^1/Series 60 5th Edition have been heavily criticized.

In November 2010, Smartphone blog *All About Symbian* criticized the performance of Symbian's default web browser and recommended the alternative browser Opera Mobile.[71] Nokia's Senior Vice President Jo Harlow promised an updated browser in the first quarter of 2011.[72]

Malware

Symbian OS was subject to a variety of viruses, the best known of which is Cabir. Usually these send themselves from phone to phone by Bluetooth. So far, none have taken advantage of any flaws in Symbian OS – instead, they have all asked the user whether they would like to install the software, with somewhat prominent warnings that it can't be trusted.

However, with a view that the average mobile phone user shouldn't have to worry about security, Symbian OS 9.x adopted a UNIX-style capability model (permissions per process, not per object). Installed software is theoretically unable to do damaging things (such as costing the user money by sending network data) without being digitally signed – thus making it traceable. Commercial developers who can afford the cost can apply to have their software signed via the Symbian Signed [24] program. Developers also have the option of self-signing their programs. However, the set of available features does not include access to Bluetooth, IrDA, GSM CellID, voice calls, GPS and few others. Some operators have opted to disable all certificates other than the Symbian Signed certificates.

Some other hostile programs are listed below, but all of them still require the input of the user to run.

- Drever.A is a malicious SIS file trojan that attempts to disable the automatic startup from Simworks and Kaspersky Symbian Anti-Virus applications.
- Locknut.B is a malicious SIS file trojan that pretends to be a patch for Symbian S60 mobile phones. When installed, it drops a binary that will crash a critical system service component. This will prevent any application from being launched in the phone.
- Mabir.A is basically Cabir with added MMS functionality. The two are written by the same author, and the code shares many similarities. It spreads using Bluetooth via the same routine as early variants of Cabir. As Mabir.A activates it will search for the first phone it finds, and starts sending copies of itself to that phone.
- Fontal.A is an SIS file trojan that installs a corrupted file which causes the phone to fail at reboot. If the user tries to reboot the infected phone, it will be permanently stick on the reboot, and cannot be used without disinfection –

that is, the use of the reformat key combination which causes the phone to lose all data. Being a trojan, Frontal cannot spread by itself – the most likely way for the user to get infected would be to acquire the file from untrusted sources, and then install it to the phone, inadvertently or otherwise.

A new form of malware threat to Symbian OS in form of 'cooked firmware' was recently demonstrated at the International Malware Conference, MalCon, December 2010, by Indian hacker Atul Alex.[73] [74]

Bypassing platform security

Symbian OS 9.x devices can be hacked to remove the platform security introduced in OS 9.1 onwards, allowing users to execute unsigned code.[75] This allows altering system files, and access to previously locked areas of the OS. The hack was criticised by Nokia for potentially increasing the threat posed by mobile viruses as unsigned code can be executed.[76]

See also

General

- EKA2 current Symbian kernel, successor of EKA1
- History of Symbian
- List of Symbian devices
- MOAP user interface
- Nokia Ovi suite
- Series60-Remote
- Nokia PC Suite, software package used to establish an interface between Nokia mobile devices and computers running Microsoft Windows operating system; not limited to Symbian
- Nokia Software Updater
- S60, Series 60, user interface used by Nokia and others
  - Web Browser for S60 default web browser
- Ovi store Nokia's application store on the Internet, not limited to Symbian
- Symbian Foundation
- Symbian Ltd.

Development-related

- Accredited Symbian Developer
- Active object (Symbian OS)
- Carbide.c++, alternative application and OS development IDE
- Cleanup stack
- P.I.P.S. Is POSIX on Symbian
- Python for S60, alternative application development language
- Qt (framework), preferred development tool, both for the OS and applications, not limited to Symbian
  - Qt Creator IDE
  - Qt Quick
  - QML, JavaScript based language

Symbian^3 EPL Source

- symbiandump [77]
- wildducks [78]

Applications

- See category:Symbian software (still very incomplete)

References

[1] Nokia and Accenture Finalize Symbian Software Development and Support Services Outsourcing Agreement (http://newsroom.accenture.com/news/nokia-and-accenture-finalize-symbian-software-development-and-support-services-outsourcing.htm)
[2] Lextrait, Vincent (January 2010). "The Programming Languages Beacon, v10.0" (http://www.lextrait.com/Vincent/implementations.html). . Retrieved 5 January 2010.
[3] Not Open Source, just Open for Business (http://symbian.nokia.com/blog/2011/04/04/not-open-source-just-open-for-business/). symbian.nokia.com (2011-04-04). Retrieved on 2011-09-25.
[4] History of Symbian
[5] Lee Williams Symbian on Intel's Atom architecture (http://web.archive.org/web/20090419214755/http://blog.symbian.org/2009/04/16/symbian-on-intels-atom/). blog.symbian.org. 16 April 2009
[6] Uikon-Eikon-Avkon-Qikon - Nokia Developer Wiki (http://www.developer.nokia.com/Community/Wiki/Uikon-Eikon-Avkon-Qikon)
[7] Lunden, Ingrid (2011-09-30). "Symbian Now Officially No Longer Under The Wing Of Nokia, 2,300 Jobs Go" (http://moconews.net/article/419-symbian-now-officially-no-longer-under-the-wing-of-nokia-2300-jobs-go/). *moconews.net*. . Retrieved 30 September 2011.
[8] Nokia announces Symbian 'Anna' update for N8, E7, C7 and C6-01; first of a series of updates (video) (http://www.engadget.com/2011/04/12/nokia-announces-symbian-anna-update-for-n8-e7-c7-and-c6-01/). Engadget. Retrieved on 2011-09-25.
[9] Nokia announces Symbian Belle alongside three new devices (http://www.engadget.com/2011/08/24/nokia-announces-symbian-belle-running-on-three-new-devices/). Engadget. Retrieved on 2011-09-25.
[10] "infoSync Interviews Nokia Nseries Executive" (http://www.infosyncworld.com/news/n/11070.html). Infosyncworld.com. 2010-06-24. . Retrieved 2010-08-12.
[11] 100 Million Club H1 2010 (http://www.visionmobile.com/blog/2010/10/smart-feature-phones-the-unbalanced-equation-100-million-club-series/). VisionMobile (2010-10-18). Retrieved on 2011-09-25.
[12] RIP: Symbian (http://www.engadget.com/2011/02/11/rip-symbian/). Engadget. Retrieved on 2011-09-25.
[13] Epstein, Zach. (2011-06-23) Symbian is officially no longer Nokia's problem (http://www.bgr.com/2011/06/23/symbian-is-officially-no-longer-nokias-problem/). Bgr.com. Retrieved on 2011-09-25.
[14] Symbian OS – one of the most successful failures in tech history (http://eu.techcrunch.com/2010/11/08/guest-post-symbian-os-one-of-the-most-successful-failures-in-tech-history/). TechCrunch.com. November 8, 2010
[15] Symbian Foundation (2010-02-04), *Symbian Completes Biggest Open Source Migration Project Ever* (http://www.symbian.org/news-and-media/2010/02/04/symbian-completes-biggest-open-source-migration-project-ever), , retrieved 2010-02-07
[16] Menezes, Gary. (2010-09-11) Symbian OS, Now Fully Open Source (http://www.watblog.com/2010/02/06/symbian-os-now-fully-open-source). Watblog.com. Retrieved on 2011-09-25.
[17] "Symbian Foundation website, members section" (http://www.symbian.org/members). .
[18] "Developer Economics 2011" (http://www.visionmobile.com/blog/2011/06/developer-economics-2011-winners-and-losers-in-the-platform-race/). .
[19] symbian-dump | Download symbian-dump software for free at. Sourceforge.net. Retrieved on 2011-09-25.
[20] symbian-incubation-projects – Symbian Incubation Projects – Google Project Hosting (http://code.google.com/p/symbian-incubation-projects/). Code.google.com. Retrieved on 2011-09-25.
[21] *Sibo3a screenshots* (http://www.guidebookgallery.org/screenshots/sibo3a), Guide Book Gallery, .
[22] Marcin Wichary. "GUIdebook - Screenshots - EPOC R5/Psion Revo" (http://www.guidebookgallery.org/screenshots/epocr5). Guidebookgallery.org. . Retrieved 2010-08-12.
[23] Solution to Nokia Slow SMS / Hang Problem (http://www.kejut.com/nokiasms)
[24] http://www.symbiansigned.com
[25] http://www.techradar.com/news/phone-and-communications/mobile-phones/nokia-n8-to-get-symbian-belle-update-in-early-2012-1046874
[26] http://conversations.nokia.com/2011/12/21/nokia-belle-coming-soon/
[27] Nokia PR (Oct 21, 2010). "Nokia further refines development strategy to unify environments for Symbian and MeeGo" (http://www.nokia.com/press/press-releases/showpressrelease?newsid=1453894). . Retrieved 2010-11-05.
[28] AllAboutSymbian (Oct 26, 2010). "The future of the Symbian platform" (http://www.allaboutsymbian.com/features/item/12223_The_future_of_the_Symbian_plat.php). . Retrieved 2010-11-05.

[29] Nokia PR (May 24, 2006). "Nokia releases 'Web Browser for S60' engine code to open source community" (http://www.nokia.com/A4136002?newsid=1052589). *press.nokia.com*. . Retrieved 2007-03-21.

[30] Browser and Maps updates for many S60 3rd Edition and S60 5th Edition phones (http://www.allaboutsymbian.com/news/item/13056_Many_S60_3rd_Edition_and_S60_5.php). All About Symbian (2011-06-29). Retrieved on 2011-09-25.

[31] "Symbian — Qt – A cross-platform application and UI framework" (http://qt.nokia.com/products/platform/symbian/). Qt.nokia.com. . Retrieved 2010-08-12.

[32] Nokia Developer (2010-06-18), *Nokia Qt SDK* (http://developer.nokia.com/Develop/Qt/), , retrieved 2012-01-20

[33] Apps:Mobile Web Apps in a Nutshell (http://www.symlab.org/wiki/index.php/Apps:Mobile_Web_Apps_in_a_Nutshell). symlab.org wiki

[34] Nokia Developer – Web (http://www.forum.nokia.com/Technology_Topics/Web_Technologies/Web_Runtime/). Forum.nokia.com. Retrieved on 2011-09-25.

[35] "Qt Labs Blogs " Nokia Qt SDK 1.0 released" (http://labs.trolltech.com/blogs/2010/06/23/nokia-qt-sdk-10-released/). Labs.trolltech.com. . Retrieved 2010-08-12.

[36] "Qt Labs Blogs " Qt Simulator is going public" (http://labs.trolltech.com/blogs/2010/05/31/qt-simulator-is-going-public/). Labs.trolltech.com. . Retrieved 2010-08-12.

[37] "Symbian developer community" (http://developer.symbian.org). Developer.symbian.org. 2010-01-27. . Retrieved 2010-08-12.

[38] http://www.oracle.com/appforge/

[39] http://www.redfivelabs.com/content/WhyNet60.aspx

[40] Tom Sutcliffe and Jason Barrie Morley Xcode Symbian support (http://symbian-xcode-plugin.tigris.org/). Symbian-xcode-plugin.tigris.org. Retrieved on 2011-09-25.

[41] "Capabilities (Symbian Signed) – Symbian Developer Community" (http://developer.symbian.org/wiki/index.php/Capabilities_(Symbian_Signed)). Developer.symbian.org. . Retrieved 2010-08-12.

[42] Nokia Developer News | Nokia Now Signing Symbian Apps for Free – Nokia Developer Blogs (http://blogs.forum.nokia.com/blog/nokia-developer-news/2010/08/16/nokia-now-signing-symbian-apps-for-free). Blogs.forum.nokia.com (2010-08-16). Retrieved on 2011-09-25.

[43] "Symbian developer community – technology domains" (http://developer.symbian.org/main/source/technology_domains/index.php). Developer.symbian.org. . Retrieved 2010-08-12.

[44] "Symbian developer community – packages" (http://developer.symbian.org/main/source/packages/index.php). Developer.symbian.org. . Retrieved 2010-08-12.

[45] "Symbian System Model – Symbian Developer Community" (http://developer.symbian.org/wiki/index.php/Symbian_System_Model). Developer.symbian.org. . Retrieved 2010-08-12.

[46] Introducing EKA2, by Jane Sales with Martin Tasker (http://media.wiley.com/product_data/excerpt/47/04700252/0470025247.pdf). (PDF) . Retrieved on 2011-09-25.

[47] Six Years of Symbian Produces 100 Models and 100 Million Shipments (http://www.thesmartpda.com/50226711/six_years_of_symbian_produces_100_models_and_100_million_shipments.php), The Smart PDA.

[48] Symbian Foundation Adds New Member, Nuance (http://news.softpedia.com/news/Symbian-Foundation-Adds-New-Member-Nuance-117209.shtml). News.softpedia.com (2009-07-21). Retrieved on 2011-09-25.

[49] http://www.symbianblogs.com/history/110-symbian-2--history.html

[50] Nokia N8 User Agent Profile (http://nds.nokia.com/uaprof/NN8-00r100-3G.xml). Nds.nokia.com (1999-02-22). Retrieved on 2011-09-25.

[51] Nokia launches mobile TV | Nokia Conversations – The official Nokia Blog (http://conversations.nokia.com/2010/09/09/nokia-launches-mobile-tv/). Conversations.nokia.com (2010-09-09). Retrieved on 2011-09-25.

[52] Obsolete (http://www.silverlight.net/getstarted/devices/symbian/). Silverlight.NET. Retrieved on 2011-09-25.

[53] Inside Symbian SQL: A Mobile Developer's Guide to SQLite |

By Ivan Litovski, Richard Maynard, 2010, page 9

[54] Help – Eclipse Platform (http://library.forum.nokia.com/index.jsp?topic=/Web_Developers_Library/GUID-B7D6EFF3-16E6-45D5-9B74-8333BA83FC7B.html). Library.forum.nokia.com. Retrieved on 2011-09-25.

[55] on your Mobile (http://www.skype.com/intl/en/get-skype/on-your-mobile/?cm_mmc=m102). Skype. Retrieved on 2011-09-25.

[56] Screenshot for Symbian OS | AntonyPranata.com 2.0 (http://www.antonypranata.com/screenshot/screenshot-symbian-os). Antonypranata.com. Retrieved on 2011-09-25.

[57] NTT DoCoMo releases S^2 devices (http://web.archive.org/web/20100824165132/http://blog.symbian.org/2010/06/01/ntt-docomo-releases-s2-devices/). blog.symbian.org 1 June 2010

[58] http://www.symbian.org/devices?manufacturer=All&platform=Symbian^2&form-factor=All&date_announced[value][year]=&date_announced[value][month] (http://www.symbian.org/devices?manufacturer=All&platform=Symbian^2&form-factor=All&date_announced[value][year]=&date_announced[value][month])

[59] "Symbian^2 platform used in eleven new models of NTT DoCoMo FOMA 3G handsets" (http://www.symbianone.com/content/view/7108/). SymbianOne. . Retrieved 2010-11-10.

[60] Samsung OMNIAHD Dazzles at Mobile World Congress with Its HD Brilliance (http://www.samsung.com/uk/news/newsPreviewRead.do?news_seq=12421). Samsung.com. Retrieved on 2011-09-25.

[61] Pettey, Christy. "Gartner Says Worldwide Mobile Device Sales to End Users Reached 1.6 Billion Units in 2010; Smartphone Sales Grew 72 Percent in 2010" (http://www.gartner.com/it/page.jsp?id=1543014). Gartner.com. . Retrieved 2011-03-10.

[62] "Majority of smart phones now have touch screens (Canalys press release: r2010021)" (http://www.canalys.com/pr/2010/r2010021.html). Canalys.com. 2010-02-08. . Retrieved 2010-08-12.

[63] "BBC News – Google Android phone shipments increase by 886%" (http://www.bbc.co.uk/news/technology-10839034). Bbc.co.uk. 2010-08-02. . Retrieved 2010-08-12.

[64] Nokia Leading Smartphone Market with 56%, While Symbian's Share of OS Market Is Set to Fall | Press Release (http://www.abiresearch.com/press/826). ABI Research. Retrieved on 2011-09-25.

[65] Gartner Says Sales of Mobile Devices in Second Quarter of 2011 Grew 16.5 Percent Year-on-Year; Smartphone Sales Grew 74 Percent (http://www.gartner.com/it/page.jsp?id=1764714). Gartner.com. Retrieved on 2011-09-25.

[66] Nokia and Microsoft enter strategic alliance on Windows Phone, Bing, Xbox Live and more (http://www.engadget.com/2011/02/11/nokia-and-microsoft-enter-strategic-alliance-on-windows-phone-b/). Engadget. Retrieved on 2011-09-25.

[67] Woods, Ben. (2010-10-01) Samsung to drop Symbian support | Wireless – CNET News (http://news.cnet.com/8301-1035_3-20018315-94.html). News.cnet.com. Retrieved on 2011-09-25.

[68] Meyer, David. (2008-11-03) Motorola ditches Symbian, announces 3,000 layoffs | Networking | ZDNet UK (http://www.zdnet.co.uk/news/networking/2008/11/03/motorola-ditches-symbian-announces-3000-layoffs-39539063/). Zdnet.co.uk. Retrieved on 2011-09-25.

[69] Mello, John P.. (2010-10-15) Sony Ditches Symbian (http://www.pcworld.com/article/208000/sony_ditches_symbian.html#tk.mod_rel). PCWorld. Retrieved on 2011-09-25.

[70] Mobile-reviews. Review of Nokia E7. http://mobile-review.com/review/nokia-e7-en.shtml

[71] Mobile browser comparison, November 2010 (http://www.allaboutsymbian.com/features/item/12323_Mobile_browser_comparison_Nove.php). Allaboutsymbian.com (2010-11-25). Retrieved on 2011-09-25.

[72] Meyer, David (November 9, 2010). "Nokia times first Symbian updates for 'early 2011'" (http://www.zdnet.co.uk/news/mobile-it/2010/11/09/nokia-times-first-symbian-updates-for-early-2011-40090806/). ZDNet UK. . Retrieved January 4, 2011.

[73] Hacker plants back door in Symbian firmware – The H Security: News and Features (http://www.h-online.com/security/news/item/Hacker-plants-back-door-in-Symbian-firmware-1149926.html). H-online.com (2010-12-08). Retrieved on 2011-09-25.

[74] Hacker Creates Modified Symbian S60 Firmware with Hidden Back Door (http://www.livehacking.com/2010/12/10/hacker-creates-modified-symbian-s60-firmware-with-hidden-back-door/). Live Hacking (2010-12-10). Retrieved on 2011-09-25.

[75] Nokia's S60 3rd Ed security has been hacked? (http://www.symbian-freak.com/news/008/03/s60_3rd_ed_has_been_hacked.htm), Symbian Freak

[76] "S60 v3 Hacking – Mission accomplished, FP1 hacked!" (http://www.symbian-freak.com/news/008/03/s60_3rd_ed_feature_pack_1_has_been_hacked.htm). Symbian Freak (2008-03-27). Retrieved on 2011-09-25.

[77] http://symbiandump.sourceforge.net/

[78] http://sourceforge.net/apps/mediawiki/wildducks/index.php?title=Main_Page

Bibliography

- Morris, Ben (22 June 2007). *The Symbian OS architecture sourcebook : design and evolution of a mobile phone OS* (http://eu.wiley.com/WileyCDA/WileyTitle/productCd-0470018461.html). John Wiley & Sons. p. 630. ISBN 0470018461.

External links

- Symbian foundation blog (which the homepage redirects to) (http://symbian.org/)
- Symbian (http://www.ohloh.net/p/symbian/analyses/latest) on Ohloh
- Symbian (http://www.dmoz.org/Computers/Mobile_Computing/Symbian/Symbian_OS/) at the Open Directory Project
- Qt developer website (http://developer.qt.nokia.com/)
- Symbian C++ developer website (http://www.developer.nokia.com/Develop/Other_Technologies/Symbian_C++/)

Article Sources and Contributors

Nokia_5500_Sport *Source*: http://en.wikipedia.org/w/index.php?title=Nokia_5500_Sport *Contributors*: Acalamari, Armando, Blakegripling ph, Chowbok, Cnilep, Dreadstar, Ed g2s, Editore99, GregorB, Gueneverey, Hetar, IceCreamAntisocial, Jamez654, Julkku, Kathzzzz, Kirosana, Lonet, MMuzammils, Nathandelaselva, Netrat, Sardanaphalus, Searchmaven, Sigurdhu, Slusk, Smooth O, TigerK 69, Ultratomio, Wibbble, Zooterkin, 27 anonymous edits

S60_(software_platform) *Source*: http://en.wikipedia.org/w/index.php?title=S60_%28software_platform%29 *Contributors*: 0xFFFF, A Man In Black, A:-)Brunuś, Abbasrazanasir, Adamhauner, Ahmed abbas helmy, Ahoerstemeier, Aishazc, Alf Boggis, Algocu, AlistairMcMillan, Andries, Andros 1337, Anonywiki, Antonypranata, Apocalypso™, Auntof6, Benjaminhill, Benlisquare, Bhimaji, Bjelleklang, Blackgadget, Blakegripling ph, Bluemoose, Brent01, Cameron Scott, Catalina-symbina, Chris the speller, Codetiger, Colenso, CommonsDelinker, CraigBox, Crialdo, CristiS, CyberSkull, Cyrius, Dale Arnett, Darlamack, Daverocks, Davorg, Dboehmer, Death Master, DireWolf, DiscoverYellow, Djr xi, Dori, Dp76764, Dreambringer, EdBever, Engelbert, EoGuy, Estoy Aquí, Fdavis99, Feezle, Findepi, Finnce, Flauto Dolce, FleetCommand, Florin92, Fram, FreewareLovers, Galoubet, Gerhman, Gmihaylo, Gomcoite, GraemeL, GregorB, Gwernol, Gzbr, Haakon, Handiplus, Harriv, Hawklord, Hedge777, IJK Principle, Ianweston, Ifad, Impi, Int19h, JJJCrom, JYi, JaGa, Jack007, Jannex, Jarvik, Jerryobject, Jinkm, Jlin17, Jodi.a.schneider, Jondel, Joshuaali, Jupter-manzana, Jusdafax, Kcome, Kcsomisetty, Klubneeka, Know seekerinsas, Kocio, Kresnas, Kurt Shaped Box, Kutu su, Leandrod, Lnichols, Lordofthe9, Lzur, MMuzammils, Ma401, Mabdul, Malikaqdas, Marketingemcc, Marlow4, Martarius, Matemaciek, Matt Crypto, Mdwh, MementoVivere, Metageek, Mobicawilmslow, Mononoke, Mortense, MrOllie, Mwtoews, MySchizoBuddy, N4931, Nabil2199, Neilc, Nopetro, Notmicro, Ohnoitsjamie, Oleksandr Kononenko, Orderud, PDD, PTSE, Petri Krohn, Philofred, Pnm, Polluks, Price1d, Prolog, R2j7, Radagast83, Rajeshontheweb, Rjanag, Rüdiger Wölk, SMC, Saarab, Sandox, Sarikoro, Scientizzle, Seth Nimbosa, Shombal, Sigmundur, Silvestre Zabala, Sinigagl, Slitchfield, Smyth, Speculatrix, Srikanthjnr, Steady.eddie, SteinbDJ, Sujith84, Tawker, Tazztone, TexasAndroid, The Seventh Taylor, Thomas Larsen, Towel401, Twp, Uker, Uruseidaimon, Utils, Vahid83, Varungargi, Vegaswikian, Vkem, Wafulz, Wehe, Wibbble, WikHead, XLoffers, Xaje, Xiyulangzi2008, YmFzZTY0, Younome, Yousaf465, Zpetro, İsByTiti, 336 anonymous edits

Binary_file *Source*: http://en.wikipedia.org/w/index.php?title=Binary_file *Contributors*: 16@r, Aitias, Andonic, Antonio Lopez, Blessed sea6, CBM, Captain-n00dle, Chandrakanthng, Chris Pickett, Colejohnson66, Comrade Tux, Danmichaelo, Davnor, Dawnseeker2000, Deconstructhis, Doom lord ki, ELeschev, EncMstr, Fuzzbox, Gilliam, HappyDog, Incnis Mrsi, Iranway, JSpung, Jeffq, Jyril, Khawaga, Kubanczyk, Kyng, LOL, Leafyplant, MER-C, Malenien, Marcos canbeiro, MaximvsDecimvs, Meesa 67, Michael Hardy, Milan Keršláger, Mwtoews, Noclevername, Paulnasca, Pengo, Peyronnin, PhuryPrime, Prashantgonarkar, RainR, Rd232, Reisio, RossPatterson, RufusDoofus, Sciurinæ, Scorchsaber, Seidenstud, Splintax, Stepa, Stephanbim, Svick, Teo64x, Thumperward, Trevbork, Widefox, Wikimoder, 95 anonymous edits

Speech_synthesis *Source*: http://en.wikipedia.org/w/index.php?title=Speech_synthesis *Contributors*: 12 Noon, 5 albert square, 67-21-48-122, 7, A3 nm, ACSE, Aaronrp, Abdull, Ace Frahm, Actam, Agoubard, Al Lemos, Altermike, Amakuha, Andreas Bischoff, AndrewHowse, Angr, Argon233, Arj, Arvindn, Auntof6, Auric, BStrategic, Back ache, Backstabb, Badly Bradley, Beetstra, Betterusername, Bigbubba954, Bikepunk2, BjarteSorensen, Blk2006, Bobbo, Bobo192, Bocharov, Bongwarrior, Bootedcat, Bovineone, Bradfuller, Brianreading, Brion VIBBER, Bumm13, CHIPSpeaking, Calltech, Calton, Caltrop, Canderra, Canis Lupus, CanisRufus, Cassowary, Chachou, Charlie danger, Chmod007, Chocolateboy, Chris the speller, Chrischris349, Chrysoula, Chuck Sirloin, ChuckOp, Codehydro, Conscious, CortlandKlein, Crystallina, Cybercobra, Dadu, Damian Yerrick, Darkspartan4121, Dave w74, David Eppstein, David Gerard, Ddp224, Dennishc, Diamondland, Discospinster, Dogman15, Doodle77, Dr.K., Drivertodriver1, Ds13, Dwiki, Dycedarg, Dylan Lake, E946, EdC, Eik Corell, Ellamosi, Escientist, Eurleif, Everyking, Flatterworld, Forthepoeple, Fr33kman, Fractaler, Ftiercel, Furrykef, GVnayR, Gaius Cornelius, Gary Cziko, Geothermal, Gerstman ny, Giftlite, Glenn, Googly2006, Grendelkhan, GreyCat, Gwalla, H, Hagiza121, Hamedkhoshakhlagh, Harryboyles, Hero10all, Heron, Hhanke, Hike395, Holizz, Husky, Instine, Intgr, Invitatious, Itman, Itzcuauhtli, JJblack, JLaTondre, Jacob Poon, Jagged 85, Jeff Henderson, Jenliv2ryd, Jerryobject, Jimfbleak, Jimich, Jlittlet, JoeSmack, Joelr31, JoergenB, JohnMRye, Johnny Au, Jonkerz, Jose Manuel Ortega Torres, Jphvs, Justin W Smith, K7aay, Kaldosh, Kate, Kayemel3, Kcordina, Kindall, Knowledgerend, Koavf, Kuszi, Kwamikagami, Kwekubo, Kwizy, KyraVixen, L736E, Ladysmithrose, Lakshmish, Lalalele, Latitude0116, Ligulem, LittleHow, Lksdfvbmwe, Lukeluke1978, Lupin, MER-C, MaGa, Mac, MaltaGC, MarcS, Marechal Ney, Marskell, Martarius, Martinevans123, Matrixbob, Matt Crypto, MatthPeder, Maury Markowitz, MaviAteş, Michal Jurosz, MonteChristof, Morfeuz, MotherFerginPrincess, Moyogo, MrTree, MuthuKutty, N1RK4UDSK714, Nahitk, Naikrovek, Nameless Voice, Nanard, NawlinWiki, Neelix, NeonMerlin, Nige7, Nlu, Nohat, Nopetro, Oemengr, Oli Filth, Outriggr, Oznull, Oznux, PAR, Palthrow, Parasane, Patrick, Paulson74, Paxse, Pb30, Pedro, Pengo, Pgillman, Pinethicket, Pjrich, PleaseStand, Politepunk, Polluks, Ppareit, PrimroseGuy, Python eggs, Quadell, R'n'B, Raffaele Megabyte, Raul654, RedWolf, Redox12, Reichenbachj, Renaudforestie, Requestion, Rich Farmbrough, Rjanag, Rjwilmsi, Rmcguire, RobertG, Roberto111199, Rogerb67, Ruud Koot, Sahassagala, Santhosh.thottingal, Satori, Savetz, Saxbryn, SchreyP, Scott Sanchez, Serezniy, Shadowjams, Shaftesbury12, Shaneymac, Shrike, SigmaEpsilon, Silas S. Brown, Simeon, Singlebarrel, Sinus Pi, SiobhanHansa, Sladen, SlowByte, Smack, Sonjaaa, Southpolekat, Speechgrl, StaticGull, Statisfactions, Stephanos Georgios John, StephenPratt, Stephenchou0722, SteveRenals, Supremeknowledge, Suruena, T33guy, TMC1221, TextToSpeech, The Founders Intent, TheFarix, Thefreethinker, Thenickdude, Thesilverbail, Thoobik, Thorenn, Thue, Thumperward, Thw416, TimMagic, Tjwood, Tlesher, Tobias Bergemann, TobyDZ, Tommy Blueseed, Tony1, Trainra, Twikir, Twistor, Twthmoses, Uluboz, ValerieChernek, Versus22, Viajero, Voxsprechen, W4HTX, W9000, Wayne Miller, Wayp123, Weheh, Wesley crossman, Wik, Wikipodium, Wingman417, Wizzard2006, Wolf grey, Xdenizen, Yamamoto Ichiro, Yiddophile, Yosri, Yrithinnd, Ziusudra, च दरका त धृतडमल, 516 anonymous edits

Motion_detector *Source*: http://en.wikipedia.org/w/index.php?title=Motion_detector *Contributors*: Alan.ca, Alvis, Alynna Kasmira, Amalas, Ammowriter, Andrew28913, Anlace, Atlant, Barti 81, Beetstra, Behnoodnoei, Benjrh, CHG, Chrisch, Clarince63, Csigabi, DARTH SIDIOUS 2, Dav!dB, Dg2006, Dhollm, Diego Moya, Dmclong, Editor randy, Electron9, Emana, Frap, Gadfium, GraemeL, Ithizar, Jessepotrie, Jncraton, Johnanth, Justvote, Kbdank71, Kku, Klaser, L Kensington, LoveGirlsUK, Lowellian, Mallenme, Neptune5000, Nmesisgeek, Ohnoitsjamie, Old Moonraker, Phil flip, Radagast83, Reswobslc, Sally May Roberts, SchuminWeb, Science3456, Screen111, Selket, Speakma, The Founders Intent, Tommy2010, Tudy77, Umesh89, Uncle G, Uncle Milty, Wmahan, Wtmitchell, Wtshymanski, Zahn, ZimZalaBim, 93 anonymous edits

Office_suite *Source*: http://en.wikipedia.org/w/index.php?title=Office_suite *Contributors*: A man alone, A2Kafir, After Midnight, Agujero Negro, Ak112358, Akohler, Akosnagy, Aljullu, AndersL, André437, Apofisu, Apudur, Arched013, Arvindkst, Ary29, Axl, Beland, Blanchardb, Borgx, Btyner, CC90, Cahk, Campoftheamericas, Catamorphism, Certes, Chadoh, Charwinger21, Chris the speller, Christopher Hill, Closedmouth, Coconut-Freak, Conradov, Crocodile Punter, Crysb, Czarkoff, DARTH SIDIOUS 2, Dadu, DanimothWiki, Darrelljon, Davidhorman, Deineka, Den fjättrade ankan, Dhshah, DisillusionedBitterAndKnackered, Dlrohrer2003, Dmitri Lytov, DragonFury, Econterms, El Pantera, Eofficesuite, Filll, Forteller, Ft93110, GCW50, GameKeeper, Gary King, Ghettoblaster, Gpd, Grandscribe, Gronky, Gurch, Gökhan, Haha169, Hawaiian717, Hede2000, Herald Alberich, Hkyeung, Horoporo, Ibraheem alex, Ilfironltd, Impaciente, Inhumandecency, Inimino, Interwsw, Itsmine, JLaTondre, JMS Old Al, Jacob Poon, JadziaMD, Jan.hajer, Jasón, Jedi1yoda1, Jerryobject, Jim1138, Jinghuliu, Jlsoaz, Joshua Issac, Juan M. Gonzalez, Jusses2, KAMiKAZOW, KDesk, KP-Adhikari, Kai445, Kalivd, Karnesky, Kieff, Kimdino, Kitamozihr, KitchM, Kotu Kubin, Koweja, Kozuch, Kuru, Lady Aleena, LaosLos, Lbecque, Lester, Leuce, Leuqarte, Liangent, Lillian Olina, Linberry, Lionel Elie Mamane, Lksajeev, Lmatt, Love2fish, Mac, Madize, Mahanga, Maharris777, Marclaporte, Marinus Bier, Marosha, Masky2, Maxberlitz, Maxeboy, Megaboz, Megasquid500, Meighan, Merbenz, Mets501, Mikemuch, Mikeprodgers, Minghong, Mnts, MonteChristof, Mr. Met 13, MrMan, Mrkahuna, Mysdaao, NCdave, NellieBly, Newfoundlander, Nigelj, Odoyle5150, Off!, Ohnoitsjamie, Outs, Paul63, Plasticup, Pmedema, Psantora, Pseudo daoist, Qwe, Raider Duck, Ramesh debata, Replysixty, Rhe br, Rjwilmsi, RobinZwama1, RockyMM, Rutje, Ryan Norton, SL93, Safuman, SamJohnston, Samboy, Samsara, Sanjiv swarup, Sasanjan, Satinyou, Sbalk, Scarlet Lioness, Scientus, Sean daj, Seidenstud, Sent NL, Septegram, Sgant, Shogunu, SiobhanHansa, Sitenl, Skybon, Smurfy, Supportreq, Swerdnaneb, Syock, TangLab, Tchannon, Techmdrn, Tesseran, The Anome, The Thing That Should Not Be, TheParanoidOne, Thearcher4, Themania, Themfromspace, Thumperward, Tigernike1, TimBowen78, Timothyhouse1, Tokek, Tuxcantfly, Tyoh, UnitedStatesian, Vespristiano, Voidvector, Voidxor, Welsh, Wikieditoroftoday, Wikipelli, Woohookitty, Ykhwong, Z10x, Zollerriia, 364 anonymous edits

List_of_Nokia_products *Source*: http://en.wikipedia.org/w/index.php?title=List_of_Nokia_products *Contributors*: -Majestic-, 159753, 6300i, A bit iffy, Abeyi76, Aeons, Ahodacsek, AimalCool, Ajcfreak, Ale85, Alex Neman, Alin.olaru, Analoguedragon, Andres, Andries, AndriuZ, Andros 1337, Anetode, Angel Virus, Apeman2001, Arendedwinter, Armando, Arun2007th, Ash73, Atamenes, Balcora, Bartekz89, Bautze, Benjamin9003, Beseven286, Bigbmc26, Bike21, Bitarin, Black Kite, Blakegripling ph, Bmwt, Bobblewik, Borgx, Braverman02, Buxtehude, CHG, CLW, CanisRufus, Cen, Chowbok, ClaretAsh, Comstock, CoolingGibbon, Coredesat, Cthegooner, Cvf-ps, D2s, DJFLEX-mk2, Daddygee, Daimore, Dancter, Darin-0, Davewho2, Dazzafar, DeLarge, Deono, Deusfaux, Dgies, Distortedvision, Djr xi, DynamicDes, E Wing, Ed g2s, Ed50, Edcolins, Egil, Ehn, Epoxed, Eurolite x3, Ex nihil, FH33333248, Fiftyquid, Filipvr, FleetCommand, Florin92, Folksong, Freakofnurture, Gjuteriet00, GoingBatty, Gordeonbleu, Gpvos, Grafen, GreatWhiteNortherner, GregorB, Gustav 7, Haakon, Hapo, Havarhen, Hgb asicwizard, Hooperbloob, Ice000000, Ihatetoregister, Illutionz, Ilyushka88, Imroy, Infantrymarine25, Inzy, Ioliver, Ironfrost, Irstu, ItsMeOrYou, JIP, JaGa, Jack007, Jamesp, Janta, Jasonon, Jawsper, Jedo1507r, Jeepday, Jekotira, Jkstark, Jmoz2989, Jni, Joolz, Joseph Solis in Australia, Josh Parris, Joshua Scott, Jpk, Juliabackhausen, K-links, KC., Karan.102, Ketiltrout, Kimchi.sg, Klubneeka, Kuuks, Kyrios320, LittleWink, Liuhb86, Lord Anubis, Luigiacruz, Lwc, M2ger, MMuzammils, Magioladitis, Marc NL, MarioV, Mariusmiti, Mark, Markus Kuhn, Mathias-S, Matthewsmith, Melaen, Miborovsky, Michael Devore, Michael Greiner, Mikeblas, Mild Bill Hiccup, Minfo, Minghong, Mlaffs, Moreati, Morton.lin, Mpj, Mr Stephen, MrHaiku, Mt7, Muhandes, Mvuijlst, Nakakapagpabagabag, Neandercris, Neo-Vortex, Netrat, Newone, Nickcerda, Nijel, Niteowlneils, Nokia releases, Nosilleg, Notmicro, Ntm, Nudecline, Obkt, Oscar, PTSE, Paige Master, Parag28, Pb30, Pboy2k5, Pengo, Peter S., Petri Krohn, PhilKnight, Philg88, Pisit2799, Pjahr, Pjb007, Pmlineditor, Pol098, Prateek.p1, Prunk, Pt, Putnik, Pwjb, R'n'B, R. fiend, RadioActive, Rap raz, Renegade00100, Retrowow, Rjwilmsi, Royan, Ruleke, Runtime, Ryanaxp, SF007, ST47, Saimhe, Samdlaw, Samoojas, Scofield 20, Scott5114, Seraphimblade, Sexie, Shake waves2001, Shandris, Shanes, Sharkface217, Shashankbhat, Shattered, Shirifan, Simon12, SimonP, Skype in Nokia, Slammer111, Slo-mo, Smooth O, Snickerdo, Snowmanradio, Stattouk, Steven9212, Strom, Stuartfang, Sunstep, Sverdrup, Tamilxbangalore, Taras, Taza insane, TelecomNut, Tide rolls, TigerK 69, Timberframe, Tohlz, Treekids, TreyGeek, Turkeyphant, Typ932, Ultravisitor, Umairanwer, Umar1996, Utkarsh apple, Vahid83, Vegaswikian, Vespristiano, Visor, Vkem, WISo, Weihao.chiu, Wibbble, Wiki Wonda, Wikibarista, Wikieditor06, Wikiliki, Wurdnurd, XLoffers, Xaje, Xezbeth, YUL89YYZ, Yamla, Yasirmturk, Ytrewq17, Zackwee, Zagothal, Zeerak88, ZeroOne, Zerofire0, Zirka, Zouzzou, 676 anonymous edits

Smartphone *Source*: http://en.wikipedia.org/w/index.php?title=Smartphone *Contributors*: 16@r, 3Coins, A Man In Black, A123a10, A8UDI, AKA MBG, ALSLUG, AVM, Aaditya 7, Acalamari, Acather96, Aclassifier, Acolin f, Adriatikus, Aeons, Ahoerstemeier, Aillema, Akuyume, Alansohn, Alex890, Alf Boggis, Alf.laylah.wa.laylah, AlistairMcMillan, Allen3, Amarendra.Avinash, AndrewSpec, Andries, Andros 1337, Angela, AniRaptor2001, Ansible, Anupam, Arab Dynasty, Arafael, Arny, Arp120, Arunsingh16, Atama, Atenyi, Atul.ecn, Avoided, BD2412, Bahaltener, Bakshi41c, Balazer, Baronnet, Beland, Bendes, Bhny, Bibijee, Biker Biker, BillHaywood, BlkStarr, Bomazi, Bovineone, Branddobbe, Brianreading, Broccoli, CLW,

CRJO-CRJO, Cactusframe, Cadiomals, Caj27, Calcwiki, CalumCook234, CanadianPenguin, Canalteen, CatherineMunro, Cellcom, ChaChaFut, Changshui88, Charivari, Chris the speller, Ciphers, Citizensmith, Cleared as filed, Closedmouth, Clovis Sangrail, Cmf, Cmlewan, Cmp101, Codetiger, Comaleaf, CommonsDelinker, Commontimect, Cool Hand Luke, Coolaaron88, CoolingGibbon, Cornea503, Cst17, Currab, Cxtom, Cyruslei, DaisyField, Dale Arnett, Dan6hell66, Daniel.finnan, DaveChild, Dcljr, Dcxf, Ddavid2005, Deane@gooroos.com, Delphii, DerBorg, Deviceapps, Diannaa, Diego Moya, DivineAlpha, Dmarquard, Dreambringer, Drobatch, E.au, EVula, EWikist, Editor182, Editrrr, Elassint, Electronicguru1, Emma23 K, Epolk, Eraserhead1, Erc, Erianna, Eshade, EugeneZelenko, Eugenia loli, Evice, Evilruletheworld96, Evolutiondb, F1MotoGPWRC, FCYTravis, Farolif, Father Goose, Fcassia, Feinoha, Feydey, Fhontoy, Flowanda, Fourdee, Furrybeagle, Fuzheado, GTBacchus, Galadh, Garant^ ^, GateKeeper, Geologyguy, Georgy90, Giftlite, Giggett, Gilliam, Ginsengbomb, Giraffedata, Glacialfox, Glennwells, Gmumru, Gogo Dodo, GoingBatty, Gold Hat, Gomm, Gookey, Gordon Ecker, GraemeL, Grahamperrin, Grajales, Graywolf, Graywolfmoon1, Gregorysalt, Greswik, Groink, Gsarwa, Guy Harris, Gwernol, HDCase, Ha us 70, Haakon, Hadiceberg, Halloween.mac, Hamiltha, Hammersoft, Hardylane, Harmil, Harryzilber, Haseo9999, Hede2000, Henrik, HereToHelp, Hervegirod, Hobartimus, Hu12, Human.v2.0, Huntersquid, HuskyMoon, HybridBoy, Hydrogen Iodide, Hydrox, Hypnotist uk, I already forgot, I, Podius, Ianb, Ianboudreault, Ianereed, Icedwater, Ignis Fatuus, Illegal Operation, Illyria05, Im Buff, Immunmotbluescreen, Imroy, Imsilly123, In2thats12, Indefatigable, Interframe, InternetMeme, Intershark, Ipsign, Irky, Ishdarian, ItsZippy, J, J.delanoy, JCDenton2052, JCrue, JNW, Jagdterrier, JamesWeb, JamieS93, Jantangring, Jeffq, Jekyllhide, Jerome Charles Potts, Jerryobject, Jfdwolff, Jim.henderson, Jimmy Bergmark, JoeSmack, John of Reading, JonHarder, Jonabbey, Jondel, JonoP, Joseph Solis in Australia, Josh Parris, Jostake, Jsherwood0, Jsribeiro, Juan M. Gonzalez, Jusdafax, Just another editor, Justin Mauger, Jwojdylo, Jæs, Kajowi, Kariteh, KarmaGeddon, Katieh5584, Kbdank71, Kcomstock, Kellen`, KelleyCook, Kentynet, Kevin Dorner, Kingpin13, Kinst, Klokbaske, KnowBuddy, Koavf, Kozuch, Kraftlos, Kresp0, Kudpung, KungFuMonkey, Kuru, Kwaichi, LaVieEntiere, Lai888, Lambtron, Larkhill97, Laurasmith70308, Lazulilasher, LeilaniLad, Lenin1991, Lester, Leszek Jańczuk, Leujohn, Level plus, Lewispb, Lifes g00d 561, Lilac Soul, Limequat, Little Mountain 5, Little Professor, Logan, Loser997, Lotje, LrdChaos, Lukan4ica, Lun Esex, Lun4tic, MER-C, MMuzammils, Mac, Mac John Concord, Magnus.de, Malik Shabazz, Mange01, Maniacgeorge, Manop, Manway, Marcisjustice, Marco.difresco, Marcus Qwertyus, Mark Kim, Marksza, Marrowmonkey, Martarius, Masgatotkaca, Mathiastck, Maury Markowitz, Maxí, Mcduck, Mckote, Mdrejhon, Mdwh, Meehawl, Meelar, Memaster3, Mephistophelian, MetaManFromTomorrow, Miaow Miaow, Mikael Häggström, Mike Dillon, Mike Linksvayer, Mikehelms, Mild Bill Hiccup, Minghong, Miquonranger03, Mix Bouda-Lycaon, Mlg07e, Moberg, ModWilliam, Monaarora84, Mononomic, Morphh, Mortense, Mosmof, Mr. Strong Bad, MrOllie, Mvjs, My76Strat, MySchizoBuddy, MyronAub, Myscrnnm, Nakakapagpabagabag, Nathan94124, Navacell, Nazrich, Ne0Freedom, Nealmcb, Neo Ogilvy, Neoarchon, Neoguy999115, Nerdeff, Netrapt, Nick Number, Nickolai 420, Night Gyr, Nightscream, Nitro.ajb, Nixdorf, Nneonneo, Nopetro, Nurg, Obey, Ohnoitsjamie, Oknazevad, OlavN, Old port, Oli Filth, Omegatron, Oontay, OpenToppedBus, P.Marlow, PS., PTSE, Parintachin, Patrick, Pats1, Patvac-chs, Paul 1953, Pbrown111, Pdahomepage, Petershank, Peyre, Phatalflaw, Phatom87, Phearson, Philip Trueman, Piano non troppo, Pingveno, Piotrek54321, Pluma, Pmarshal, Pmlineditor, Pol098, Pol430, Poor Yorick, Posix memalign, Prathameshsasane, Prillen, Pvanheus, Pyfan, Quebec99, Querencia, R'n'B, RA0808, Radiier, Radon210, Rafael.sp, Ramu50, RedWolf, Reliablesoft, Repetition, Res2216firestar, ReverseEngineered, Reyk, Rgreed, Rhobite, Riadlem, Rich Farmbrough, Richardguk, Richiekim, Riki, Rollins83, Ron2, Ronz, Ruchir257, Rursus, SF007, SPQRobin, Sadegh87, Sainath468, Samad120, Samwb123, Sandstein, Sapibobo, Sayid98, Schrödinger's Cake, ScottyBerg, Scrtcwlvl, Sdfisher, Sdrazfar, Sean13zz, Sebastian Mandrean, Sebindcruz, Secretsorry, Serg3d2, Sfacets, Sfan00 IMG, Sfsmartp, Shalroth, Shawnc, Sheehan, Signalhead, Siliconov, Singerdg1, SirJibby, Skatebiker, Skintigh, Skittle, Slitchfield, Smart1954, Smmgeek, Snigbrook, Soliloquial, SpaniardGR, Spellmaster, Starofwonder, Stephen B Streater, Stephen Turner, Stjson, SuperHamster, Surenkarapetyan, Suwatest, Swoof, TVS99, Taka76, Takamaxa, Taketa, Tangent1000, Taskinen, TastyPoutine, Tatterfly, Technopat, Tesi1700, Tfgbd, ThaWhistle, ThaddeusB, The Pink Oboe, The Thing That Should Not Be, The Wild Falcon, The-apathy, Theaveng, Thestick, Thiled, Thisma, Thumperward, Ticell, Tide rolls, Tikru8, TimSE, Tklaer, Tmuller2, Tobias Bergemann, Togopogo, Tokyogamer, Tomas.turek18, Tombomp, Tommy2010, Tommytentimes, Tony1, Toussaint, Tpbradbury, Trasz, Treekids, Tuju, UCLATre, UnrealG, Urod, Uzume, Vegaswikian, Verkinto, Verne Equinox, Versus22, Vkem, Vrenator, Waqas Hussain, Westie4321, WhisperToMe, Whkoh, WiZZLa, Wideangle, Wiki admi, Wikimhb, WikipedianYknOK, Wikipelli, Winged-stone, Wireless Buddy, Wknight94, Woohookitty, XGrape, Xajel, Xrobertcmx, Yean3d, Youxiarock, Yug, Yunshui, Zach.vega1, Zanter, ZimZalaBim, ZipoB, Zpetro, 1099 anonymous edits

Symbian *Source*: http://en.wikipedia.org/w/index.php?title=Symbian *Contributors*: AJ-India, Ahmadadam96, Airplaneman, Andries, Anujgupta2 979, Apdevries, Arbitrarily0, Armando, Arnaud92, Atlan, Aymen ka, B0o-supermario, BD2412, Bajjibala, Belastro, Bpringlemeir, Brianreading, BrideOfKripkenstein, C933103, Cameron Scott, Capdor, CapitalLetterBeginning, Ceancata, Cherkash, Chris the speller, Clsin, CommonsDelinker, Danhash, Daniel.Cardenas, DarwinPeacock, Davidjcole, Digitalsurgeon, DmitryKo, Editor182, Edward, Elephant in a tornado, Empty Buffer, Erianna, ErkinBatu, Favonian, Fences and windows, Ferrero777, Fetchcomms, Ffaarrooqq, FinnsDeal, FireyFly, FleetCommand, Florin92, Freddiegjertsen, Fru1tbat, Gene91, Gerhman, Gmihaylo, Gryllida, Gsarwa, Haakon, Hersfold, Hoshie, Hu12, Hydrox, IJK Principle, Imcdnzl, Intractable, IsmaelLuceno, Iuhkjhk87y678, JHP, JaGa, Jaizovic, Jayantanth, Jeffrey04, Jerryobject, Jfourgeaud, Jfromcanada, Jondoelocksmith, Jwoodger, Khanayub1986, Kinema, Kkm010, Knkw, Knurdtech, Koavf, Krashlandon, KrisBogdanov, Krischik, Kyrios320, LemChops, Lester, Ligertail, LokiClock, Lori-m, Ltomuta, Magister Mathematicae, MargaritMargaryan, Martarius, Materialscientist, Mattkap2, Mdwh, Mephistophelian, Micahblitz, MichaelBillington, Minikola, Mnia786, Mrprajesh, Muhandes, Mushroom9, Myscrnnm, Narasimhanator, Nate1481, NotWith, Notedgrant, OlEnglish, Parthibls, Pnm, Pol098, Polluks, Ponydepression, Quietbritishjim, R'n'B, Rabit gti, Reboot, Sainath468, Seth Nimbosa, Sharcho, Shrish, Shīrudou ōru, Smileverse, SoWhy, Soulxlight, Soumilj, Stichbury, Surya Prakash.S.A., Svick, Thumperward, Timotheus Canens, Tirim4, Tomjenkins52, Traveler100, Trusilver, Truth Lover80, UKER, Ufim, VMlemon, Vervamon, VijayPadiyar, Vishnunj, Wattyirl, Weeder 001, Why vincent, Yavoh, Yorxs, Zackaback, Zagothal, ZeroOne, Zhongle, ZirconiumTwice, Zvrkljati, 244 anonymous edits

Image Sources, Licenses and Contributors

File:Nokia5500frontview.jpg *Source*: http://en.wikipedia.org/w/index.php?title=File:Nokia5500frontview.jpg *License*: unknown *Contributors*:

Image:Nokia N8 Symbian Belle Screenshot.jpg *Source*: http://en.wikipedia.org/w/index.php?title=File:Nokia_N8_Symbian_Belle_Screenshot.jpg *License*: unknown *Contributors*: Future Perfect at Sunrise, Gerhman

Image:S60 3rd Edition from N73.jpg *Source*: http://en.wikipedia.org/w/index.php?title=File:S60_3rd_Edition_from_N73.jpg *License*: unknown *Contributors*: Melesse, Sujith84

File:5800 idle screen.png *Source*: http://en.wikipedia.org/w/index.php?title=File:5800_idle_screen.png *License*: unknown *Contributors*: Estoy Aquí, Hydrox, JaGa, McLoaf, 1 anonymous edits

File:Wikipedia's W.svg *Source*: http://en.wikipedia.org/w/index.php?title=File:Wikipedia's_W.svg *License*: unknown *Contributors*: Jonathan Hoefler

Image:Wikipedia favicon hexdump.svg *Source*: http://en.wikipedia.org/w/index.php?title=File:Wikipedia_favicon_hexdump.svg *License*: unknown *Contributors*: User:Mwtoews

Image:Stephen Hawking.StarChild.jpg *Source*: http://en.wikipedia.org/w/index.php?title=File:Stephen_Hawking.StarChild.jpg *License*: unknown *Contributors*: NASA

File:TTS System.svg *Source*: http://en.wikipedia.org/w/index.php?title=File:TTS_System.svg *License*: unknown *Contributors*: Andy0101 (talk). Original uploader was Andy0101 at en.wikipedia

Image:Motion detector.jpg *Source*: http://en.wikipedia.org/w/index.php?title=File:Motion_detector.jpg *License*: unknown *Contributors*: User:CHG

File:LibreOffice Writer 3 3 2 en.png *Source*: http://en.wikipedia.org/w/index.php?title=File:LibreOffice_Writer_3_3_2_en.png *License*: unknown *Contributors*: User:Akosnagy

File:Office_suite_ja.png *Source*: http://en.wikipedia.org/w/index.php?title=File:Office_suite_ja.png *License*: unknown *Contributors*: Akabe, Gareth

File:Nokia1100.jpg *Source*: http://en.wikipedia.org/w/index.php?title=File:Nokia1100.jpg *License*: unknown *Contributors*: MacUsr

File:Nokia1108.jpg *Source*: http://en.wikipedia.org/w/index.php?title=File:Nokia1108.jpg *License*: unknown *Contributors*: User:Diego Grez

File:Nokia 1110 DG 01.jpg *Source*: http://en.wikipedia.org/w/index.php?title=File:Nokia_1110_DG_01.jpg *License*: unknown *Contributors*: Original uploader was DogGunn at en.wikipedia

File:Nokia 1112.jpg *Source*: http://en.wikipedia.org/w/index.php?title=File:Nokia_1112.jpg *License*: unknown *Contributors*: User:Pooh01

File:Nokia 1200 Macic7 01.jpg *Source*: http://en.wikipedia.org/w/index.php?title=File:Nokia_1200_Macic7_01.jpg *License*: unknown *Contributors*: Original uploader was Macic7 at en.wikipedia

File:Nokia1202arab.jpg *Source*: http://en.wikipedia.org/w/index.php?title=File:Nokia1202arab.jpg *License*: unknown *Contributors*: User:Saiht

File:Nokia 1208 ubt.JPG *Source*: http://en.wikipedia.org/w/index.php?title=File:Nokia_1208_ubt.JPG *License*: unknown *Contributors*: © 2010 by

File:Nokia1209.jpg *Source*: http://en.wikipedia.org/w/index.php?title=File:Nokia1209.jpg *License*: unknown *Contributors*: Basvde, Nokia 1209 Phone

File:Nokia1600.png *Source*: http://en.wikipedia.org/w/index.php?title=File:Nokia1600.png *License*: unknown *Contributors*: User:Dzoker, User:TheAdam0s

File:Nokia 1610.jpg *Source*: http://en.wikipedia.org/w/index.php?title=File:Nokia_1610.jpg *License*: unknown *Contributors*: User:Erik Baas

File:Nokia 1661 front.JPG *Source*: http://en.wikipedia.org/w/index.php?title=File:Nokia_1661_front.JPG *License*: unknown *Contributors*: User:Cherryguy93

File:Nokia 1662.jpg *Source*: http://en.wikipedia.org/w/index.php?title=File:Nokia_1662.jpg *License*: unknown *Contributors*: User:Ytrewq17

File:Nokia 1800.jpg *Source*: http://en.wikipedia.org/w/index.php?title=File:Nokia_1800.jpg *License*: unknown *Contributors*: Pjahr

File:Nokia e50.jpg *Source*: http://en.wikipedia.org/w/index.php?title=File:Nokia_e50.jpg *License*: unknown *Contributors*: User:Asimzb

File:Nokia E51 Black.jpg *Source*: http://en.wikipedia.org/w/index.php?title=File:Nokia_E51_Black.jpg *License*: unknown *Contributors*: Original uploader was Feci1024 at en.wikipedia

File:Nokia E52 2.jpg *Source*: http://en.wikipedia.org/w/index.php?title=File:Nokia_E52_2.jpg *License*: unknown *Contributors*: User:Asimzb

File:Nokia E55 01.jpg *Source*: http://en.wikipedia.org/w/index.php?title=File:Nokia_E55_01.jpg *License*: unknown *Contributors*: James Nash

File:Nokia E60.jpg *Source*: http://en.wikipedia.org/w/index.php?title=File:Nokia_E60.jpg *License*: unknown *Contributors*: MalcolmScott (talk)

File:Nokia E61 pl.png *Source*: http://en.wikipedia.org/w/index.php?title=File:Nokia_E61_pl.png *License*: unknown *Contributors*: User:Kangel, User:TheAdam0s

File:Nokia E62.jpg *Source*: http://en.wikipedia.org/w/index.php?title=File:Nokia_E62.jpg *License*: unknown *Contributors*: Roland Tanglao from Vancouver, Canada

File:Nokie E63 View.png *Source*: http://en.wikipedia.org/w/index.php?title=File:Nokie_E63_View.png *License*: unknown *Contributors*: Jack Dyson

File:Nokia e65.jpg *Source*: http://en.wikipedia.org/w/index.php?title=File:Nokia_e65.jpg *License*: unknown *Contributors*: Original uploader was Gyung at en.wikipedia

File:E66 vertikalne.jpg *Source*: http://en.wikipedia.org/w/index.php?title=File:E66_vertikalne.jpg *License*: unknown *Contributors*: User:El Carlos, User:Miraceti

File:Nokia e70.jpg *Source*: http://en.wikipedia.org/w/index.php?title=File:Nokia_e70.jpg *License*: unknown *Contributors*: Bror Heinola

File:Nokia E71.JPG *Source*: http://en.wikipedia.org/w/index.php?title=File:Nokia_E71.JPG *License*: unknown *Contributors*: User:MenoK

File:Nokia e72-1.jpg *Source*: http://en.wikipedia.org/w/index.php?title=File:Nokia_e72-1.jpg *License*: unknown *Contributors*: User:Stocksy

File:Nokia E75 (open)-3284596583 063f9ce1d0 o.jpg *Source*: http://en.wikipedia.org/w/index.php?title=File:Nokia_E75_(open)-3284596583_063f9ce1d0_o.jpg *License*: unknown *Contributors*: http://www.flickr.com/photos/james_nash/

File:Nokia-E90-3.jpg *Source*: http://en.wikipedia.org/w/index.php?title=File:Nokia-E90-3.jpg *License*: unknown *Contributors*: Original uploader was Weberflo at de.wikipedia

File:Group of smartphones.jpg *Source*: http://en.wikipedia.org/w/index.php?title=File:Group_of_smartphones.jpg *License*: unknown *Contributors*: gillyberlin

File:IBM SImon in charging station.png *Source*: http://en.wikipedia.org/w/index.php?title=File:IBM_SImon_in_charging_station.png *License*: unknown *Contributors*: User:Bcos47

File:Nokia 9210.jpg *Source*: http://en.wikipedia.org/w/index.php?title=File:Nokia_9210.jpg *License*: unknown *Contributors*: J-P Kärnä

File:Htc Touch Pro2 Georgy.JPG *Source*: http://en.wikipedia.org/w/index.php?title=File:Htc_Touch_Pro2_Georgy.JPG *License*: unknown *Contributors*: User:Georgy90

File:Original iPhone docked.jpg *Source*: http://en.wikipedia.org/w/index.php?title=File:Original_iPhone_docked.jpg *License*: unknown *Contributors*: Andrew from London, UK

File:Galaxy Nexus smartphone.jpg *Source*: http://en.wikipedia.org/w/index.php?title=File:Galaxy_Nexus_smartphone.jpg *License*: unknown *Contributors*: Faramarz, MB-one, SF007, 1 anonymous edits

File:Global Mobile Applications Store Revenue.svg *Source*: http://en.wikipedia.org/w/index.php?title=File:Global_Mobile_Applications_Store_Revenue.svg *License*: unknown *Contributors*: User:MySchizoBuddy

File:Smartphone share current.png *Source*: http://en.wikipedia.org/w/index.php?title=File:Smartphone_share_current.png *License*: unknown *Contributors*: -- Eraserhead1 <talk> 12:49, 3 March 2010 (UTC) Graph created by myself. Original uploader was Eraserhead1 at en.wikipedia

File:Symbian logo 4.svg *Source*: http://en.wikipedia.org/w/index.php?title=File:Symbian_logo_4.svg *License*: unknown *Contributors*: Avicennasis, Editor182, Fetchcomms, Lester, Sfan00 IMG

CPSIA information can be obtained at www.ICGtesting.com
Printed in the USA
LVOW050759260612
287678LV00003B/21/P